NORTH JERSEY BEER

A BREWING HISTORY FROM PRINCETON TO SPARTA

CHRIS MORRIS

AMERICAN PALATE

Published by American Palate
A Division of The History Press
Charleston, SC 29403
www.historypress.net

Cover photo of the George Washington Bridge by John A. Dryzga.

First published 2015

ISBN 978.1.54021.331.0

Library of Congress Control Number: 2015931113

Notice: The information in this book is true and complete to the best of our knowledge. It is offered without guarantee on the part of the author or The History Press. The author and The History Press disclaim all liability in connection with the use of this book.

CONTENTS

ACKNOWLEDGEMENTS

Beer is good.
−Tim Morris (my dad)

Writing a book is hard. And it takes more than one person to do it, even if only one name appears on the front cover. This is my first book, so I leaned heavily on a lot of people, all of whom I feel the need to thank.

First and foremost, there's my family and friends, all of whom had to listen to me talk about this book nonstop during the four months it took me to write it. My parents, Tim and Jackie, were especially helpful during this process. One day, my dad came to me and said, "I want you to quote me in your book." "What's the quote?" I asked. "Beer is good." There's no one I enjoy sharing a beer with more than my dad, so I thought, as a thank-you, I'd give him his quote. I also have to thank my brother, Tim, and sister, Kelly. When I brewed my first beer, Tim was right there with me. He's been there for most brew days since. Every year, the two of us have a good old-fashioned brew-off where we put our beers head to head. I'm still contesting the results (his wife, Ashley, always says mine is better), but if I really did lose, there's no one I'd rather lose to. We hope to start a brewery together one day, and my mom is already asking bartenders if they have us on tap. Finally, she won't get any credit on the cover, but Kelly was my copy editor for this book, and without her, I would have handed in a book with some embarrassing typos. Thanks for your help, Kelly. And to the rest of my very close-knit extended family, thank you.

Acknowledgements

I also have a lot of friends who were supportive—in some cases overly supportive—during this process. Without that support, I don't know that this book would have ever gotten written. I couldn't possibly name them all, so I won't even try. Thank you guys, for everything.

I also have to thank Whitney Landis, my commissioning editor, for proposing this book and seeing me through it until the end. Readers should know that she did a lot more than her job description says she does.

I'd also like to thank all the people I've worked with and learned from over the years, from my time at The College of New Jersey and Rutgers to everyone from the *Star-Ledger*.

This book took a lot of effort to write, not just by me but by other people whose research was utilized throughout. You may never read this book or know that that you contributed to it, but to Lew Bryson, Mark Haynie, Ed Taggert, Marc Mappen and Michael Pellegrino, thank you.

I also have to thank the people this book is about: the great people behind craft beer in New Jersey. People like Dave Hoffman, Greg Zaccardi, Bob Fuchs, Chris Walsh, Glenn Bernabeo, Brendan O'Neil, Dave Manka, Mike Cerami, Rick Reed, Brian Boak, Bob Olson and Andrew Maiorana have all contributed immensely to making the North Jersey beer scene as great as it is. A special thanks goes to Paul Silverman of New Jersey Beer Company, who put up with a number of annoying e-mails and phone calls from me. Even though they're in South Jersey, another thanks goes to Augie Carton (Carton Brewing) and Michael Kane (Kane Brewing) for all the fun chats.

Finally, I'd like to thank Bruce Springsteen and the E Street Band, whose music got me through long car rides to breweries and hours and hours of writing.

INTRODUCTION

When Whitney Landis of The History Press first proposed this book, I was excited. As a New Jersey beer writer, I've spent years trying to help advance the craft beer industry in my favorite state. That said, I also thought I had my work cut out for me. I was right. At the time I was writing this book, I had been a homebrewer for almost five years and had been writing about New Jersey beer for the *Star-Ledger*, New Jersey's largest newspaper, for about three years. I had come to learn an awful lot about beer, specifically in New Jersey. Still, I expected this project to be a challenge. I figured I'd start by splitting it up into two parts: before Prohibition and after. Initially, that's what I did, but it didn't last. No, a book on beer in the Garden State would need more than just two parts.

The first part of this book—"Beer, Liberty and Revolution"—is relatively short but important nonetheless. Chapter 1 begins with what has been dubbed "The Craft Beer Revolution" and gives a history of craft beer since the mid-1960s, when Fritz Maytag purchased Anchor Brewing in San Francisco. Chapter 2 then gives a brief introduction to what beer is, how it's made and why it's such a fun thing to learn about. Chapter 3, the final chapter in Part I, tells a little bit on the history of beer in the founding of America. Did you know that the *Mayflower* stopped at Plymouth Rock because it was out of beer?

Part II, "The History of Beer in North Jersey," essentially has three sections: before Prohibition, during Prohibition and after Prohibition. Chapter 4 begins with New Jersey's earliest breweries and brings us to the early 1900s, when Newark was a brewing mecca and the big brewers of the Garden State—Ballantine, Kruegers, Feigenspan, Hensler, Wiedenmayer and others—were recognized around the country. That all came to an end on January 16, 1920, when Prohibition went into effect.

Growing up in New Jersey, I knew that the history of beer during Prohibition would cover more than just a few pages. I was, after all, well aware of the amount of organized crime that occurred during this time. Still, I didn't anticipate dedicating as much time and effort to it as it ultimately proved it needed. The stories of breweries during Prohibition, raids by federal agents and murders to quiet witnesses are plenty. Add in the stories of the gangsters who controlled it all—like Longy Zwillman, "Waxey Gordon," Max Hassel and "Big Maxie" Greenberg—and you have several books' worth of material. Chapters 5 and 6 cover some of the good ones but not all of them. If you find these chapters interesting, I suggest reading *Prohibition Gangsters: The Rise and Fall of a Bad Generation* by Marc Mappen; *Bootlegger: Max Hassel, the Millionaire Newsboy* by Ed Taggert; and *Gangster No. 2: Longy Zwillman, the Man Who Invented Organized Crime* by Mark Stuart. They will at least get you started. Chapter 7 finishes off the old history of beer in northern New Jersey, taking us from the end of Prohibition in 1933 to 1994, when only one New Jersey Brewery was left standing: Anheuser-Busch.

Part III, "The Craft Beer Revolution," finishes off the book by bringing us from 1994 to where we are today. Chapter 8 covers the years between 1994 and 2000, beginning with Climax Brewing, the first modern craft brewer in New Jersey, and ending with Cricket Hill. Chapter 9 then takes us from 2005 to 2014, which saw breweries like BOAKS, New Jersey Beer Company and Bolero Snort start brewing in the Garden State. If you'd like to learn more about these breweries, there are two things you can do. First, you can read Lew Bryson and Mark Haynie's *New Jersey Breweries*. Or even better, you can go visit them yourself! They're all in business as of this writing. Finally, Chapter 10 concludes the story of beer in northern New Jersey.

One thing should be noted as you begin this book. Even if you aren't from or have never lived in New Jersey, you are probably well aware of the heated debates that take place when discussing New Jersey geography. When I was an undergraduate at The College of New Jersey, there was one question for which everyone had a passionate answer: does Central Jersey exist? As a graduate student at Rutgers University (the State University of New Jersey), I learned that this debate is just one of many. Therefore, I find it might be helpful to explain the borders of what this book considers "North Jersey." While I am passionate about the existence of Central Jersey (I've lived there my entire life), the history of beer in the Garden State is best tackled in two parts—one on beer in the north half and one on beer in the southern half. Therefore, in this book, if it's north of I-195, it's North Jersey; if it's south of I-195, it's South Jersey. Essentially, this book, for the sake of keeping things neat, counts Central Jersey as North and the Jersey Shore as South.

With that said, you can now enjoy this book. I hope you find it interesting and learn while reading it as much as I did writing it. New Jersey breweries are brewing some of the finest craft beer in the world—I would know, as I've been writing about them for years. One of my goals in writing this book—besides telling the fun history of beer in North Jersey—is to promote our amazing breweries, which in my opinion don't get enough recognition. So, let me leave you with one final note: go visit them. Even if it's just one. If you live in New Jersey, chances are there's a brewery producing world-class beer within a few miles of your house. Even if you don't live in New Jersey (I'm sorry to hear that), you probably have one nearby. Go visit it. Go visit a few. They create jobs, give business to local farms and donate portions of proceeds from their beer to charity, and they're making a truly great product.

PART I

BEER, LIBERTY AND REVOLUTION

CHAPTER 1
THE CRAFT BEER
REVOLUTION

This history of beer in the United States is one of ups and downs. Prior to Prohibition—which prohibited the manufacture, transportation or sale of beer over .5 percent alcohol in the United States from 1920 to 1933—the brewing industry thrived, reaching a high of 4,131 breweries in 1873 that has yet to be matched since. Prohibition took its toll not only on the breweries whose businesses were affected but also on the restaurants, taverns, farmers, shippers and other industries that relied on the brewing industry. New Jersey—one of the few states that had refused to ratify the Eighteenth Amendment but had to obey it anyway—had 51 breweries at the time of Prohibition alone. Only a fraction survived Prohibition, either by changing their businesses or by illegally selling beer, and none of those was operating in a similar capacity afterward.[1]

Thirteen years after the Eighteenth Amendment went into effect, the Twenty-first Amendment repealed it, making full-strength beer legal yet again. But Prohibition's effects were still evident; when it ended, only three hundred breweries began brewing again, eight hundred fewer than were brewing just a decade earlier.[2]

Following World War II, the brewing industry underwent a decades-long period of consolidation: the big national brands like Budweiser, Miller, Coors and Pabst all got bigger, while craft breweries were forced to shut down. Bland light beer became the norm. By the 1960s, there was only one craft brewery: Anchor Brewing.

Things began to change for the better in 1965, when Frederick Louis "Fritz" Maytag III, grandson of Frederick Louis Maytag, founder of the Maytag Washing Machine Company, moved to San Francisco, as recollected in a story told in Tom Acitelli's *The Audacity of Hops: The History of America's Craft Beer Revolution*. After graduating with his degree in American literature from Stanford and doing some graduate work in Japan, Maytag returned to the West Coast at the ripe age of twenty-five to figure out his next steps.

One warm day in August 1965, he paid a visit to a restaurant he frequented called the Old Spaghetti Factory on Green Street in the city's North Beach neighborhood. As he did so often, he ordered his go-to beer, none other than Anchor Brewing's Anchor Steam beer. It was then that the restaurant's owner, Fred Huh, made his way over to Maytag.

"Fritz, have you ever been to the brewery?" He asked.

"No."

"You ought to see it," Huh replied. "It's closing in a day or two, and you ought to see it. You'd like it."[3] Fred Kuh didn't know it, but he had just started the craft beer revolution.

The next day, Fritz Maytag made his way to the brewery on Eighth and Brannan Streets. After sitting in the taproom with Lawrence Steese—the owner of the failing brewery—for about an hour, he fell in love with it and purchased a 51 percent stake. Maytag kept Steese on to help brew and run the business, since he had little knowledge of the industry and the brewing process, and he became head salesman. As he walked the streets of San Francisco trying to sell Anchor beers, bar and restaurant owners thought he was a little crazy; many didn't even believe that Anchor was still brewing. Maytag began to plan out the future of his business—how he would market it, produce it and distribute it. As it turns out, he was planning the beginning of a much larger craft beer movement.

Over the following years, Maytag had to reinvent Anchor's beer, and Anchor Steam became *the* California common beer, a relative of steam beer that dated back to the California Gold Rush, when Anchor was born to Gottlieb Brekle. Within the next decade, Anchor had increased capacity from six hundred barrels per year to six thousand (in the United States, one barrel is equal to thirty-one gallons; in the United Kingdom, a barrel is forty-three gallons), and distribution, which used to include only a few San Francisco taverns, expanded to other parts of California along with Arizona, Nevada and Colorado.[4]

With a population demanding more flavorful beer, as was evidenced by Anchor's revival, a wave of craft breweries soon emerged. In 1976, New

Albion Brewing Company became the first microbrewery of the modern era. A year later, Michael Jackson published *The World Guide to Beer*, which has since been translated into more than ten languages and is still considered the most influential book on craft beer. In 1978, the federal ban on homebrewing was repealed, allowing the art of brewing to become a (legal) hobby for millions. This allowed homebrewers to brew beer styles that weren't available anymore, expanding the palate of beer drinkers. It also allowed Ken Grossman to open the Home Brew Shop, a homebrew supply store in Chico, California. In 1980, he shifted gears and founded Sierra Nevada, which would lead the charge for the craft beer industry over the next few decades.

In the two decades following New Albion, some of the greatest craft beer pioneers would turn their hobbies into businesses. In Washington, Redhook opened in 1981, and Grant's Brewery Pub, later renamed Yakima Brewing & Malting Company, began brewing beer and serving food in 1982, becoming the first ever brewpub; in Michigan, Bell's Brewery opened in Kalamazoo in 1983, as did California's Hopland Brewery (now Mendocino Brewery), followed by Lagunitas in 1993 and Stone and Firestone Walker in 1996; in Massachusetts, the Boston Beer Company, better known by the brand name Samuel Adams, opened in 1984; in New York, Brooklyn Brewery opened in 1987; in Oregon, Deschuetes Brewery opened in 1988; in Colorado, New Belgium opened in 1991; and in Delaware, Sam Calagione started Dogfish Head in 1995. Back then, they were all little-known pioneers. Today, they're some of the most recognizable names in the craft beer industry.[5]

In 1965, though, there was just one. Anchor stood alone. As of the time of this writing, there are more than three thousand craft breweries in the United States, with another two thousand in the planning stages. In 1965, just a handful of national and regional breweries dominated the market; if it wasn't Budweiser, Coors, Miller or Pabst, you probably wouldn't have known it existed. Today, the majority of Americans live within ten miles of a local brewery.[6]

Craft beer has grown by more than just quantity, though. In the times of the national brands, mass-produced pilsners and similar light beers took up the majority of store shelves. As craft beer has grown, so has the American palate. Dozens of different styles now make up the market, from simple lagers to dark and roasty stouts, from brown ales to amber lagers and from wheat beers to pumpkin ales. If you think of an ingredient, chances are its been used to brew. In 2012, Wynkoop Brewing, established in 1988 as another early pioneer, brewed Rocky Mountain Oyster Stout, a stout brewed with roasted barley, Styrian Goldings hops and freshly sliced and roasted bull testicles. Not to be outdone, Stone followed by brewing a beer that used yeast found in its brewmaster John Maier's beard. These are just two examples of how far beer has come, although they are certainly two of the more extreme ones.[7]

Every day, breweries are crafting world-class beer. While some are inspired by their beards, others are inspired by more "normal" things. Take New Jersey's own Augie Carton of Atlantic Highlands' Carton Brewing, a brewmaster who's become known for creating beers inspired by food. Carton's GORP, for example, is short for "Gold Old Raisins and Peanuts," the trail mix Augie and his cousin and business partner Chris Carton used to eat while hiking. The beer takes a robust porter, adds peanuts and chocolates and reminds its drinkers of simpler times. Or take Carton's Panzanella, a beer brewed with "malts chosen to play croutons, summit hops chosen for their garlic/onion pungency" and "dry hopped with tomatoes and cucumbers evoking that classic midsummer salad."[8]

It's hard to deny the strides craft beer has made since Fritz Maytag first took a stroll down San Francisco's Eighth Street. Quantity has certainly grown at exponential rates, but so has craft beer's uniqueness, its innovation and, most importantly, its quality.

The history of beer in the United States is one of ups and downs. Right now, we're on the up.

CHAPTER 2
THE ART AND SCIENCE
OF BEER

How man discovered beer can only be speculated. The leading theory, though, is that it was an accident, albeit a happy one. It is believed that a supply of wild grain, collected and stored to be used as food, was rained on. As the water sat, it warmed up, causing the grain-water to turn into wort, which fermented from wild yeast in the air, creating beer. They drank it, and they liked it.

Dr. Soloman Katz argued that they liked it so much that early nomads abandoned their traveling ways about nine thousand years ago and settled down so that they could farm the ingredients necessary for beer, which became one of man's most important sources for nutrients—mainly protein, which unfermented grain didn't have. Once settled in one place, civilization began to form and, after that, states, governments and all the institutions that now make up our everyday lives.

Beer became an important part of newly formed civilizations. It was brewed by ancient Babylonians; the builders of the Egyptian pyramids were paid in part with beer, and pharaohs used to appoint a "Royal Chief Beer Inspector"; and the early Romans even honored barley by engraving it on their gold and silver coins.[9]

So, what exactly is beer? In its most general definition, beer is an alcoholic drink made from fermented cereal grains. These grains—such as barley, wheat, corn and rice—have starches made of sugar molecules in them that are used as food reserves to feed the grain as it grows. Through a process known as malting, the sugar-releasing process is started, resulting in malted grain. These malted grains are then steeped in hot water—usually about 150 degrees—which activates malt enzymes that convert the starches into fermentable sugars. (Unfermentable dextrin sugars are also produced. Since they don't ferment, these sugars add sweetness to the final product.) The product of the mash is called wort—a thick, sweet liquid—which is then separated from the spent grain through a process called "lautering." The grain is then rinsed through a process called "sparging" to collect any residual sugars.

The wort is then boiled—usually for sixty or ninety minutes, although it can be longer—in a large pot or kettle, during which hops are added at various increments. Hops are the female flowers of the *Humulus lupulus*, or hop plant, a vine-like plant that grows up to twenty feet tall. The flowers contain two acids known as alpha acids and beta acids.[10]

Alpha acids are important for two reasons. First, they have antibacterial properties that help prevent contamination and spoilage in beer. Their use as a preservative is actually how hops were first introduced into beer. Interestingly enough, they also have a relaxing, calming quality about them due to a chemical called dimethylvinyl carbinol—a pillow full of hops was once used as a treatment for insomnia.[11] The second reason alpha acids are important is that, through the boiling process, they form a compound called iso-humulone, which adds bitterness to the beer.

Beta acids, on the other hand, do not form iso-humulone during the boil and therefore offer no bitterness to the beer. Instead, beta acids are responsible for the aroma of the beer.

As a general rule, the longer hops are in the boil, the more flavor and aroma gets burned off, and only the bitterness remains; conversely, hops that are in the boil for a shorter time have less time to isomerize and therefore add less bitterness, while the aroma will remain intact. It should come as no surprise, then, that hops that are high in alpha acids are generally used early in the boil and are called bittering hops, while hops that are high in beta acids are added later in the process and are called aroma hops. While this is true, it certainly doesn't limit any type of hop to certain uses. Brewers often use hops that are high in alpha acids as late additions, since their aromatic qualities can still be desirable, and vice-versa. It's helpful to think of hops as spices that are added to beer to season it with flavor. Different types of hops containing different amounts of alpha and beta acids and having different taste and aroma characteristics contribute different properties to the beer. Just as chefs have creative freedom to use different combinations of spices in their foods, brewers use different combinations of hops.

A bowl of hop pellets sits next to a freshly tapped beer. *Chris Morris.*

After all hop additions have been added to the wort during the boil process, it is cooled; brewing yeast, *Saccharomyces cerevisiae*, is then added. When oxygen isn't present, yeast lives and grows through a process called fermentation, which is exactly what happens in beer. Here, yeast eats the fermentable sugars created during the mash and expels ethyl alcohol and carbon dioxide. It also produces other compounds familiar to beer drinkers, such as esters (fruity notes), phenols (spicy notes) and diacetyl (buttery notes).

While there are many types of yeasts, all of which ferment differently and produce different characteristics, there are two main categories: ale and lager yeast. Ale yeasts are top-fermenting and work at warmer temperatures and therefore tend to produce more fruitlike notes. While there are dozens of styles of ale beers, some of the most popular ones are pale ales and India pale ales (Sierra Nevada Pale Ale, Dogfish Head 60 Minute IPA and Carton 077XX Double IPA), blonde ales (Victory Summer Love, Flying Fish Farmhouse Summer Ale and River Horse Summer Blonde Ale), porters and stouts (Founders Porter and Guinness Stout), barleywines (Anchor Old Foghorn and Heavy Seas Below Decks) and most Belgian ales (Trappist Westvleteren 12 and Ommegang Three Philosophers).

Lager yeasts, on the other hand, are bottom-fermenting. They ferment at colder temperatures, and the resulting product is a cleaner, crisper and less fruity beer. They also produce undesirable sulfur compounds, so lagers have to be conditioned, or lagered, for several weeks at near-freezing temperatures. During this time, the yeast consumes these unsavory compounds, leaving the clean and crisp taste the brewer intended. Again, there are dozens of styles of lagers, but among them are pilsners (Beck's, Warsteiner and Sixpoint The Crisp), Märzen/Oktoberfest (Spaten Oktoberfestbier, Hacker-Pschorr Oktoberfest-Märzen and Samuel Adams Octoberfest), bocks and doppelbocks (Ayinger Celebrator, Spaten Optimator and Tröegs Treoegenator), Munich helles (New Belgium Summer Helles and Cricket Hill East Coast Lager) and adjunct lagers (Budweiser, Pabst Blue Ribbon and Miller).

As mentioned, fermentation involves the output of carbon dioxide. While many beers are force carbonated—a process by which beer is artificially carbonated using tanks of carbon dioxide—some are naturally carbonated, or bottle-conditioned. This is especially true for homebrewers, who don't have the means to force carbonate their beer, although many breweries, such as Sierra Nevada, also do it because it's said to give better flavor and mouthfeel. In bottle conditioning, a small amount of fermentable sugar is added back into fermented beer before it is bottled. The yeast, which is still living in the beer (while many breweries filter yeast out of the beer to make it clearer, this isn't done when bottle conditioning), ferments again. Through this fermentation, more carbon dioxide is created and (if the bottle is properly sealed) absorbed into the liquid, resulting in carbonated beer.

Beer, Liberty and Revolution

While yeast is now recognized as a vital ingredient in beer—beer wouldn't exist without it—that wasn't always the case. As mentioned earlier, beer is believed to have been discovered by accident; wild yeast living naturally in the air spontaneously fermented the wort that was created when grains collected for food sat in warm water. Back then, and until the late seventeenth century, brewers didn't know that yeast existed. In fact, the German Purity Law of 1516, known as the *Reinheitsgebot*, originally listed the only allowable ingredients in beer as water, barley and hops. It wasn't until the 1860s, when Louis Pasteur discovered yeast and its role in fermentation, that the law was amended to include yeast.[12]

Sam Calagione and the brewery he founded and owns, Dogfish Head, have become famous not only for their modern beer creations—like the popular 60, 90 and 120 Minute IPAs—but for their historic beers as well. Working with Dr. Patrick McGovern, an expert on ancient fermented beverages, Sam and his team have recreated several "ancient" beers whose recipes were resurrected from the dead. Among them is Midas Touch, a beer based on ancient drinks found in King Midas's 700 BC tomb.

In 2010, Dogfish Head released another in its "Ancient Ale" series: Ta Henket, which is ancient Egyptian for "bread beer." Based on hieroglyphics and remnants found in the tomb of Pharaoh Scorpion I, a monarch who lived more than five thousand years ago, Ta Henket is brewed with an ancient form of wheat and hearth-baked bread, chamomile, doum-palm fruit and the za'atar spices, which include savory, thyme and coriander. To replicate the yeast used in the ancient beer, Calagione traveled to Cairo. There, he planted sugar-filled petri dishes to catch the wild yeast in the air, which is possibly a descendant of the wild yeast that fermented the pharaoh's beer.[13]

While all four core ingredients are equally important in the brewing process, one is underappreciated: water. There's a reason, after all, why so many breweries operate where there is a known high-quality water supply, in places like Burton-on-Trent, England; Bend, Oregon; and, of course, New Jersey (okay, New York, too). The composition and flavor of a brewery's water source have significant effects on the product it's able to create. If the water is too "hard," meaning it's high in calcium and magnesium, a brewer may have a hard time brewing lagers, too "soft" and it may have a hard time brewing pale ales and bitter beers.

With water making up between 90 percent and 95 percent of beer, pure taste is a factor as well. Luckily, New Jersey and New York have excellent water. For the same reason that our pizza and bagels are so famous, brewers like Augie Carton of Carton Brewing say that New Jersey water adds something special.

While the *Reinheitsgebot* may only allow four (originally three) ingredients to be used in brewing—water, barley, hops and yeast—American craft beer, more than any other, has embraced innovation in brewing, and that includes the use of new and interesting ingredients. Dozens of fruits and vegetables—including pumpkin, apple, tomato, blueberry, blackberry, chili pepper, cherry, apricot, raspberry, peach, agave and others—have become popular ingredients alongside spices like tea, juniper, ginger, nutmeg, cinnamon and seeds of paradise. And creative brewers are using new ingredients every day.[14]

CHAPTER 3
GIVE ME BEER AND
GIVE ME LIBERTY

J ust as the story of beer dates back some ten thousand years, the story of beer in America dates back some time as well, long before the United States existed as a country. In fifteenth-century Europe, streams and lakes were contaminated, causing Europeans to avoid drinking water. Beer, however, was nontoxic because it had been boiled and contained alcohol, so it acted as a safer alternative to the contaminated water. It was also simple to make from local ingredients, plus it had some nice side effects that people enjoyed.[15]

It should come as no surprise, then, that as Europeans began traveling to the New World in the 1500s, they brought kegs of beer with them for their long, treacherous journey. In fact, not only did the Pilgrims on the *Mayflower* have beer aboard the ship, but it also played a major part in why they settled at Plymouth Rock. The Pilgrims actually had plans to sail farther south, where the weather was warmer and less dangerous. But according to the diaries of Pilgrims, they decided to settle at Plymouth Rock because they were running low on supplies, specifically beer. William Bradford, one of the men on the *Mayflower*, related on December 19, 1620, just why they considered settling sooner than planned: "We had yet some beer, butter, flesh and other victuals left, which would quickly be all gone; and then we should have nothing to comfort us." After exploring the area around Cape Cod, the time came for the Pilgrims to decide on where they would settle. On this, Bradford wrote, "We could not now take time for further search or consideration, our victuals being much spent, especially our beer." Nathaniel Philbrick, in *Mayflower: A Story of Courage, Community, and War*, further explained that after exploring the area, the Pilgrims found "several freshwater springs along the high banks of the brook that bubbled with 'as good water as can be drunk'—an increasingly important consideration now that they were forced to ration what remained of the beer."[16]

Once settled, the Pilgrims wasted no time: they built a brewery that first winter, one of the first things built in the New World. To them, beer was more than just a safe drink, it was a comfort they brought with them from the Old World. Through battling tough weather conditions, widespread disease, food shortages and homesickness, it proved to be a nice comfort to have.

As settlements began to form throughout the New World, breweries were often centerpieces of towns. While homebrewing was common, some breweries produced enough beer to sell to the public, so long as they were properly licensed. It hadn't taken long to set up a system to regulate the brewing industry: records from November 1637 show laws that fined "anyone brewing beer, malt, or other drink, without a license." Breweries also became an important selling point in colonial real estate. An ad in the *New York Gazette* in December 1752 advertised a New Brunswick, New Jersey farm for sale: "There is also a large new Brew-House, 60 Feet long, and 38 wide, with a new Copper, containing 22 Barrels, with all the Utensils proper for Brewing." Another ad in the September 12, 1754 *Pennsylvania Journal* sought buyers for a house in "Trentontown" (now Trenton); the house included "a new stone arched vault, sufficient to hold one hundred barrels of beer."[17]

Beer also played an important role in the American Revolution. Many of America's Founding Fathers were brewers; George Washington even had a brewery at Mount Vernon. Samuel Adams was also involved in brewing, as his grandfather owned a malt house on Purchase Street in Boston, and when he passed away in 1748, Adams took over. John Hancock, who owned a shipping business, was also a businessman involved in the beer industry. When the English Crown enacted the Stamp Act as a way to collect money so it could pay off debt accrued while fighting France and New World rebels, both Hancock and Adams were affected. The Stamp Act, passed in 1765, put a tax on goods made in the colonies, although the colonies had no say in this.[18]

Both Hancock and Adams organized the resistance movement—known as the Sons of Liberty—against the oppressive laws, and they often did so in pubs, especially the Black Horse Tavern in Winchester, Massachusetts. The movement led to a widescale boycott of anything taxable under the Stamp Act, which resulted in huge financial losses for the British, who repealed the Stamp Act in 1768 and replaced it with the Townsend Acts, which instead taxed imported goods. This provided a great opportunity for local business, whose products were no longer among those that were taxed, and breweries were no exception.[19]

Beer and taverns continued to play important roles in the Revolution. From 1763 to 1776, Benjamin Franklin's son, William, served as the colonial governor of New Jersey and was loyal to the Crown. In the spring of 1765, the Mutiny Acts required the colonies to provide royal troops with specific items, among them rations of beer. The Colonial Assembly in New Jersey, while loyal, was not in favor of this, and Franklin was caught in the middle. He eventually landed on the side of the Assembly and helped pass a law that allowed brewers and taverns to deny free beer to soldiers.[20]

Beer played an important role in the Revolutionary War itself as well. The Founding Fathers drank beer as they planned the Boston Tea Party and even as they debated the Declaration of Independence. It was used at recruitment events, and at George Washington's urging, it was legislated by the Continental Congress that soldiers receive a ration of one quart per day.

Even after independence had been won, beer continued to play an important role. New York City served as the United States' first capital from 1785 to 1790, and Fraunces Tavern, a place George Washington often visited, served as home to the Departments of State, Treasury and War.[21]

PART II

THE HISTORY OF BEER IN NORTH JERSEY

CHAPTER 4
BREWING BEFORE PROHIBITION

1641–1920

If you walk down Washington Street in Hoboken, there's one thing you'll see a lot of: bars. It's fitting, really, because this is where beer in New Jersey all started, back in 1641. In that year, Hoboken was leased to a Dutch immigrant named Aert Teunissen Van Patten for twelve years. There, Van Patten owned a farmhouse and a brew house in the north end of the city, making him the first brewer in the state. According to the genealogical record of his family, however, Aert died at a young age. It is believed that Leni-Lenape Native Americans killed Van Patten's livestock and burned down his house before killing him in 1643, just two years after he leased the land.[22]

And so the story of New Jersey's first brewery isn't a happy one. But Jersey brewing carried on, thanks in large part to the state's agricultural potential and fresh water supply. Throughout the next two hundred years, records show breweries operating in Burlington, New Brunswick, Trenton, Elizabeth, Mount Holly, Lamberton, Newark, Hackettstown, Weehawken and Hoboken, although not much is known about them. In the Salem colony alone, Thompson's, Nicholson's, Morris's and Abbot's were all well-known breweries at the time. In the late 1800s, the state saw an explosion in beer production. By 1900, fifty-one breweries were producing 2.5 million barrels per year, up from fifty-eight breweries producing just over 500,000 barrels per year in 1879. Although there were breweries in the south, the majority were in the north, with nearly half in Newark alone. The city was an optimal spot due to the fresh water supply and train routes to the New York market.

Today, not including the beer produced by Anheuser-Busch, New Jersey breweries produce about fifty thousand barrels of craft beer per year, just one-fiftieth of what was being produced at the turn of the twentieth century. Today, New Jersey ranks thirty-fourth in craft beer production; in 1900, it ranked seventh. Most of that came out of Newark and its surrounding cities, although New Brunswick, with its prime location near the Raritan and the King's Highway into New York was also a major brewing center. Through the mid-1800s up to Prohibition, North Jersey was a brewing powerhouse, and it all started with John N. Cumming, whose brewery would later become the giant that was Ballantine.[23]

BALLANTINE BREWERY | NEWARK

Although Ballantine—arguably New Jersey's most important brewery and one of the most important in the country—wasn't founded until 1840, the story of Ballantine dates back to 1805, when General John N. Cumming founded his brewery at High and Orange Street in Newark. Newark Brewery sold porter, single ale and double ale in bottles and kegs until it was sold in 1832. An ad in an 1831 issue of the *Albany Evening Journal* read:

> *Valuable Brewery at Auction. Will be sold at Public Auction, on the premises, on Saturday, the 3d of September next, at 3 o'clock P.M....on High-Street, in the flourishing Town of Newark, N.J. generally known as "Cumming's Brewery."...One hundred and thirty bushels of Malt have usually been brewed at a time, and with little alteration can be increased to a much larger quantity. A never failing stream of pure spring water is conducted into the building, and furnishes a constant supply in all parts of the Brewery and Premises. Upon the premises are also a Brick Office, Stables, extensive Sheds, Out Houses...about two acres of land, which can be sold with or without the Brewery. The Hogsheads, Barrels, Casks, Kegs, Bags, etc., etc., belonging to the establishment, can be had at a fair valuation...This brewery has never been able to supply the demands of the town and surrounding country. As proof of this several thousand barrels of Ale are imported to this market yearly, from Albany, Troy, Lansingburg, and other places...The Morris Canal being now in operation, and in the immediate vicinity, will afford many advantages to the establishment, not before possessed, in transporting Beer into the interior of this State, Pennsylvania, etc....*[24]

The brewery was bought by Robert Morton in 1832, who would later lease it to Thain and Collins in 1838. Two years later, in 1840, New Jersey's brewing landscape would be forever changed.[25]

PETER BALLANTINE WAS BORN on November 16, 1791, in Ayrshire, Scotland. He moved to Albany, New York, in 1820, where he began brewing, eventually opening his own brewery in 1833. In 1840, he decided to move closer to the booming New York City market, settling with his wife and kids in Newark and taking over the brewery that once belonged to General John N. Cummings along with his partner, Erastus Patterson, forming the Patterson and Ballantine Brewing Company. That partnership lasted only five years, at which point Ballantine bought out Patterson and went into business by himself. In 1850, he built a new brewery on the Passaic River, and in 1857,

A coaster from Ballantine Ale and Beer, Newark. *Bart Solenthaler, Bart & Co.*

he brought his sons—Peter H., John H. and Robert F.—into business with him, forming P. Ballantine & Sons.

By 1877, P. Ballantine & Sons was the fourth-largest brewery in the country, producing about 107,000 barrels per year. Two years later, in 1879, Ballantine used the famous three-ring symbol—"Purity, Body and Flavor"—for the first time. At the time, Ballantine was the only brewery in the top twenty that exclusively brewed ales, but that changed in 1879, when it purchased the Schalk Brewery and began brewing lagers to accommodate the changing tastes that came about with the influx of immigrants from Central Europe. By 1880, Ballantine brewing plants sat on twelve acres at Ferry and Freeman Streets.[26]

In 1882, Peter's oldest son, Peter H., passed away after becoming sick while on a trip to Europe. Peter Ballantine died soon after, on January 23, 1883, at ninety-one years of age. In addition to starting the great Ballantine brewery, he had served as a director of both the Newark City Bank and the Fireman's Insurance Company and was well known for his charity, largely to St. Michael's Hospital of Newark. Upon his death, the brewery was passed to his other son John and, finally, to his youngest son, Robert. Still, the brewery flourished. By 1895, Ballantine was producing half a million barrels per year and was the fifth-largest brewer in the country behind Pabst, Anheuser-Busch, Jos. Schiltz and Ehret Brewery.[27]

Peter H. Ballantine's daughter (Peter Ballantine's granddaughter), Sara Linen Ballantine, was born in Newark on March 5, 1858. In 1881, she married George Griswold Frelinghuysen. When Ballantine's last son, Robert, died in 1905, Frelinghuysen took control of the brewer, the first non-Ballantine to run the business since its founding fifty years earlier. The brewery continued to thrive until 1920, when Prohibition forced it to change its business model. It wasn't done brewing forever, though.[28]

KRUEGER'S BREWERY | NEWARK

While Krueger's Brewery began in 1858, its roots go back to 1851, when Louis Adam and J. Braun came together to form a new brewery. Braun, however, died before the brewery was completed, so Adam formed a new partnership with John Laible to form Laible & Adam, which began brewing 1,200 barrels that year.

Later that year, Laible wrote to his sister in Germany, asking her to send her son, Gottfried Krueger, then sixteen years old, to help at the brewery. Gottfried arrived in 1852 and began working at the brewery as an apprentice to his uncle. He continued working there until 1855, when Laible and Adam ended the partnership. Adams took over the brewery, and Laible went into business with Krueger in 1858, building a new business that Krueger would head until 1865, when Krueger partnered with Gottlieb Hill to purchase Louis Adam's brewery and began brewing under the Hill & Krueger banner.

Hill & Krueger was met with enormous success. Within a year, capacity had increased to four thousand barrels per year. That quickly grew to five thousand barrels, and by 1875, the brewery was producing twenty-five thousand barrels per year. That same year, Gottlieb Hill began experiencing health problems, and on February 16, 1875, Gottfried Krueger became the sole owner of the brewery, which now brewed as G. Krueger Brewing Company.[29]

In 1882, Krueger allowed his brewery to join a syndicate that, along with the breweries of Peter Hauck and Anton Hupfel, became the United States Brewing Company, although Krueger continued to manage his brewery. With the money he gained from the sale, Krueger was able to purchase a majority stake in Lyon & Sons Brewing. In 1908, an English syndicate of breweries began to decline, and the investors decided to sell. United States Brewing Company, now headed by Krueger, purchased it. The firm now owned the breweries of Krueger, Hauck and Hupfel, as well as Trefz and Albany Breweries, and was producing more than 500,000 barrels per year. Krueger also owned Lyon & Sons and a large interest in the Home, Eagle and Union Breweries.

Krueger, like many successful businessmen at the time, was very active in the community. In addition to owning his breweries, he served as the president of the German Savings Bank (later the United States Savings Bank), vice-president of the State Banking Company, president of the Krueger Hygiene Ice Company and director of the Union National Bank and Federal Trust Company and was once the president of both the New York Brewers' Association and the New Jersey Brewers Association. He was also involved in progressive political movements and served as a country freeholder, a member of the state legislature and a member of the Democratic State Committee's finance committee. He also served as a democratic elector in the 1880 and 1884 presidential elections and was an elector-at-large for the 1888 election.[30]

For several years, Krueger's brewery was forced to run without Krueger at the helm. In May 1914, Krueger and his wife, Bertha, sailed to Germany,

Krueger's was one of the largest breweries in New Jersey both before and after Prohibition. *Art LaComb.*

where they were both originally from. When World War I began mid-trip, the two were forced to remain in Germany. They would eventually make it back to New Jersey.

Like all breweries, Krueger's was affected by Prohibition in 1920. While the brewery would brew again, Gottfried wasn't a part of it—he died in 1926.[31]

Joseph Hensler Brewing Company | Newark

The brewery of Joseph Hensler was one of New Jersey's earliest success stories. The brewery, which Joseph Hensler founded in 1860 as a small firm, was in business for nearly one hundred years, closing in 1958. Hensler was a well-educated brewer, having worked at Lorenz and Jacquillard before going into business for himself. He realized early on how important it was to have the most up-to-date equipment, especially since his brewery largely produced lager beers, which required refrigeration to keep the beer at cooler temperatures during fermentation and conditioning (or "lagering").

By 1912, the brewery was made up of six buildings, all multiple stories. It produced nearly 200,000 barrels of beer per year and employed about

Hensler…it's a whale of a beer! *Bart Solenthaler, Bart & Co.*

two hundred people, all of whom were well paid. Joseph's son Joseph Jr. took over the brewery after his father but passed away in 1908, after which his brother, Adolph F. Hensler, took control. During Prohibition, Hensler Brewing began making soda and cereal beverage (also known as "near beer") under the banner of the Jos. Hensler Co. and ran a cold storage plant for fresh and frozen fish under the name of the Federal Food Fish Products Co.[32]

LYON & SONS BREWERY | NEWARK

The Newark Business Directory of 1858 lists a business at 28–30 South Canal Street in Newark, near current-day Broad Street, belonging to Jacob Leonhart. However, two years later, the 1860 directory shows that address belonging to the Rumpf & Frielinghaus Brewing Company.

Daniel Frielinghaus was born in Germany on March 10, 1823, the second of eleven children. Sometime around 1853–54, he came to the United States, signing his Declaration of Intent to become a United States citizen on October 9, 1854. His brother, Robert, who came to the United States in August 1854, signed his Declaration of Intent on the same day. The manifest of the ship on which Robert traveled, *Antwerp*, lists him as a twenty-year-old brewer.

In November 1859, Daniel became a naturalized citizen, and records from the next year show him as the co-owner of Rumpf & Frielinghaus. The business employed eight men until it was purchased by D.M. Lyon in 1864. In 1867, William H. Lyon joined the firm, and his brother, C.D. Lyon, joined in 1882. In 1884, Lyon & Sons was producing thirty thousand barrels per year. According to *Newark, New Jersey's Greatest Manufacturing Centre, Illustrated*, Lyon & Sons' most popular beers were its Export Edelwild, Kaiser Brau Lager, XX Ale, XXX Ale and porter. It was later purchased by Gottfried Krueger and became part of the Krueger empire.[33]

Newark city directories show Daniel Frielinghaus as owning another brewery, Traudt & Frielinghaus, with partner F.A. Traudt on the corner of Springfield Avenue and Magnolia Street from 1867 to 1869. Records from Orange Township directories show him running yet another brewery at White and Jefferson Streets in Orange, New Jersey, from 1872 to 1873. He passed away on August 23, 1873, in Newark.[34]

LEMBECK & BETZ EAGLE BREWING COMPANY | JERSEY CITY

Henry Lembeck was born on April 8, 1826, in Osterwick, Germany. He had planned on following in his father's footsteps to become a carpenter but was drafted into the German army in 1846, a year before the country's revolution, which he appeared to support. Following a furlough in March 1849, he didn't return to his regiment and instead headed west for the United States.

In his first few years in the United States, he lived in New York City, working as a carpenter. He later bought and ran a successful grocery store before joining John F. Betz as a salesman for his New York brewery. In 1969, the two went into business together, forming the Lembeck & Betz Eagle Brewing Company in Jersey City.[35]

The use of "Eagle" in the name is an interesting story in itself. Even before meeting Henry Lembeck, John Betz came from a brewing family; his brother-in-law was D.G. Yuengling, owner of America's oldest brewery,

Lembeck & Betz Eagle Brewery, Jersey City. *Library of Congress.*

which is still in operation. Betz actually learned the craft of brewing at the Yuengling Brewery in the mid-1800s alongside Henry Clausen, William Woerz and George W. Robinson, all of whom went on to start their own successful breweries. (Robinson went on to become brewmaster of Albany Brewing Company, one of the companies that was absorbed by Krueger's.)

The original name of Yuengling's brewery was Eagle Brewery (the Eagle image is still a part of Yuengling's packaging). At twenty-one years of age, Betz had completed his apprenticeship at Yuengling and returned to Europe, where he was born, to continue developing his brewing skills. When he returned to the United States, he opened a brewery on West Forty-fourth Street in New York City in 1853, naming it Eagle Brewery in tribute to his time learning under David Yuengling. He went on to open another in Philadelphia 1867, which he named after himself, before going into business with Henry Lembeck, again using the Eagle name in tribute to his time with Yuengling. He later expanded with more breweries in New York and Philadelphia.

With Betz's extensive background in brewing and Lembeck's business acumen, Lembeck & Betz Eagle was immediately successful. This success grew even greater when, in the 1880s, Lembeck noted the declining popularity of American ales. In 1889, they added to the brewery and began producing lager. A malt house in Watkinds, New York, was also a part of the firm. At its peak, the brewery produced 250,000 barrels per year and was the fourth largest in the state.

Both Lembeck and Betz became well known outside the beer industry. Betz amassed a large fortune thanks to his success in the beer industry and became active in real estate, eventually becoming one of the wealthiest men in Philadelphia. In about 1880, he built a new home, a 364-acre estate that took on the nickname "Betzwood." According to Mark A. Noon's *Yuengling: A History of America's Oldest Brewery*, "he became well known among European royalty and was granted two private audiences with Pope Leo XIII."[36]

Lembeck, too, was an active member of his permanent community, Jersey City. He was one of the founders of the Greenville Banking and Trust Company, was once the vice-president of the Third National Bank of Jersey City and served as a director for the Hudson Real Estate Company. He also owned a lot of land in Greenville, a section of Jersey City, some of which he donated to Columbia Park. He also built houses throughout the city until a dispute with the city caused him to stop.

Until the time of his death on July 25, 1904, Henry Lembeck was the president of Lembeck & Betz Eagle. Following his death, his sons,

Lembeck & Betz Eagle Brewery. *Library of Congress.*

Gustav and Otto, took control of the brewery. Like many breweries, Lembeck and Betz Eagle did not survive Prohibition—it was sold and was later used for refrigeration. In 1984, the National Register of Historic Places designated the area the Lembeck & Betz Eagle Brewing Company District.[37]

The Hauck Brewery | The Harrison Brewery | The Peter Doelger Brewery | Harrison

The story of the Peter Doegler Brewery dates back to the mid-1800s, starting with the Hauck family. Peter Hauck was born in Klin Munster, Germany, and immigrated to New York in 1844. His father owned a brewery in New York, but the business was later moved when Peter relocated to Harrison, New Jersey, in 1869. The business was originally known as Hauck & Kaufman, but in 1881, Hauck took over and the name was changed. By 1900, the brewery was producing 100,000 barrels per year and was well known for its Hauck's Special, Hauck's Extra and Golden Brew beers, all of which showcased a large *H* on the label. The brewery would later join Krueger and Hupfel to become the United States Brewing Company, which lasted until Prohibition.[38]

One Grade - One Quality

PETER DOELGER

First Prize

BOTTLED BEER

WE brew but one grade of beer—have done so for over 50 years. There is no scattering of efforts to different standards of quality. Every man devotes his entire skill and labors to the production of this one grade. Uniform quality must be the result of this concentrated effort.

Choicest Barley Malt, Rice and Saazer Hops, and a brewery that is a model in equipment and sanitation are the other foundation stones of Peter Doelger quality.

Have a case delivered to your home.

"BEER is the PURE TEMPERANCE drink"

An ad for Peter Doelger beer that appeared in a 1916 theater magazine. *Bob Haefner.*

Meanwhile, the Doelger family was becoming a well-known family of brewers in New York. Not unlike Peter Hauck, Joseph Doelger was born in Germany, immigrating to New York in 1843, just a year before Hauck. Doelger was originally employed as a barrel-maker, a vital employee of any brewery at the time. He took three years to learn the business and, in 1846, opened his own brewery.

His first brewery was on East Third Street, although that location didn't last long. His early success forced him to change locations several times, as high demand required to him to increase his capacity several times over the next decade. His success also allowed him to pay the passage for his brother, Peter, to come to the United States in 1850. In 1859, Peter opened his own brewery and then expanded in 1863.

In 1882, Joseph Doelger died, and his sons, Jacob and Anthony, renamed the company Jos. Doelger's Sons. Around the same time, Peter ran into issues with the labor movement after an accident at his brewery caused the deaths of four people. Despite a boycott, Doelger's success continued—by 1895, he was the eleventh-biggest brewer in the country. The company would survive Prohibition, although Peter wasn't alive to see it, passing away in 1912 at eighty years old. His son Peter would take control of the company.

In 1917, Peter Hauck's oldest grandson, Karl Bissel, married Phobe Doelger. This link proved valuable for the Doelger family, who took over the Hauck family brewery and then the Harrison Brewery in 1936, three years after Prohibition ended.[39]

C. TREFZ BREWING COMPANY | NEWARK

Charles Trefz was born on June 18, 1867, in New York City. Shortly after he was born, in 1869, his family moved to Newark, where his father, Christopher Trefz, founded Christopher Trefz Brewery. After attending the Military Academy in College Point, New York and the Heights Academy in Short Hills, New Jersey, Charles went to study at Newark's New Jersey Business College. He would go on to take over the brewery, located on Rankin Street between South Orange Avenue and Springfield Avenue, and it became one of the largest breweries in New Jersey.[40]

Despite both the brewery's and Charles Trefz's success—Charles went on to become a member of the New Jersey State Assembly—both are remembered largely for an 1889 accident at the brewery. When vats on the third story holding thousands of gallons of beer caused the floor to collapse, a "river of Jersey beer" was sent rushing down the street:

> *Charles Trefz's brewery…was partially wrecked yesterday by a remarkable accident. The resting and fermenting departments of the brewery are in a three-story brick building facing on Rankin-street. It contained a score or more of great vats filled with thousands of gallons of beer…Between 12 and 1 o'clock yesterday the neighborhood was startled by a terrific explosion in this building. The heavy walls trembled as though shaken by an earthquake. Then beer began to gush in streams from the rear windows, and it kept gushing into the street until it rose above the level of the sidewalks and poured into the basements of the neighboring houses. The atmosphere became suddenly so charged with ammonia that people withdrew from the streets.*

While no lives were lost, the damage was estimated at about $90,000. The brewery was sold to Krueger's in 1920.[41]

CHRISTIAN FEIGENSPAN BREWING COMPANY | NEWARK

The Christian Feigenspan Brewing Company opened in 1969 at what is now Raymond Avenue in Newark. To differentiate itself from the other large breweries in Newark at the time, its slogan was "P.O.N.," or "Pride of Newark," and those letters were lit on the Feigenspan brewery

Feigenspan P.O.N.—the "Pride of Newark." *Bart Solenthaler, Bart & Co.*

throughout the Depression until the brewery was bought by Ballantine in 1944.[42]

The brewery was run by a father-and-son team who brewed award-winning beer—their export beer won a silver medal at the 1877 Paris Exposition. When the father, Christopher, died in 1899, Christopher W. took control of the brewery. He went on to serve as the president of the United States Brewers Association and was active in the City of Newark administration, serving as the president of the Commercial Casualty Insurance Company of Newark and as a member of the "Committee of 100," which planned the 1916 celebration of Newark's 250th anniversary. He died on February 7, 1939.[43]

George W. Wiedenmayer Brewing Company | Newark

George W. Wiedenmayer was born in Newark on April 28, 1848. His father was a Newark brewer as well, opening a brewery in 1858. While the brewery closed in 1880, it allowed George to learn to brew. He worked with his father until 1877, at which point he moved to New Brunswick. He was there for two years before returning to Newark, where he founded the George W. Wiedenmayer Brewery in 1880.[44]

The brewery reached peak production at seventy-five thousand barrels per year, making it one of the larger breweries in the state. Wiedenmayer was successful outside the industry as well. He owned two steamboat lines in Newark, and like his father, who was a former mayor of Newark, he was active in politics: he was a member and one-time president of the Newark Common Council and served in the New Jersey Assembly. He passed away on September 5, 1909.[45]

Prohibition caused the company to turn into an ice cream business in 1920. The Wiedenmayer's sold the company to Foremost Dairies in the 1950s.[46]

Peter Breidt Brewery Company | Elizabeth

Like many of America's early brewers, Peter Breidt was born in Germany, on September 26, 1845. Breidt lived in Germany until 1857, at which point he was brought to the Newark by his uncle, Herr Gottlieb Schmalz. Breidt spent his first few years working various jobs before going to work with Joseph Hensler as a bookkeeper for his brewery. He worked for Hensler for four years before joining Christian Trefz, another brewer, with whom Breidt learned the brewing business. After five years, Breidt moved on to yet another brewery, Feigenspan Brewing Company, where he worked for three years.

Following his time at Feigenspan, Breidt moved to New York City, where he ran a company that exported malt beverages to the south. This business venture wasn't met with much success, however, and he sold the company at a loss.

Breidt returned to New Jersey shortly after, this time relocating to Elizabeth, where he partnered with William Laible to form Laible & Breidt in 1882 after purchasing the Eller and Bayer Company at 900 Pearl Street; it had opened in 1864 as the first brewery in Elizabeth. Less than a year into their partnership, Laible died, and Breidt took sole ownership, changing the

Breidt's Beer and Ales. *Bart Solenthaler, Bart & Co.*

name to the Breidt Brewery before incorporating as Peter Breidt Brewery Company. The brewery was met with huge success and grew larger by the year, becoming one of the largest breweries in New Jersey before Prohibition went into effect.[47]

Peter Breidt became famous name through northern New Jersey, not only because of his beer but because of his civic involvement as well. He was a member of the Free and Accepted Masons, the Benevolent and Protective Order of Elks and the Maennerchor and Liederkranz Societies, and he gave generously to many charities. He passed away on May 10, 1914. The company operated until Prohibition, at which point it shut down. Today, the site is home to Elizabeth High School.[48]

The Paterson Consolidated Brewing Company | Paterson

The Paterson Consolidated Brewing Company was a conglomerate that came about in 1890 with the union of four breweries: Braun Brewery, Sprattler & Mennell, Graham Brewery and the Burton Brewery.

The Braun Brewery

The Braun Brewery was founded in 1855 by Christian Braun at Braun and Marshall Streets in Paterson. The brewery was met with success and continued until 1870, when Braun leased it to Sprattler & Mennel. When Braun died in 1876, his sons, Christian and Louis, took over the brewery when Sprattler & Mennel moved to a new facility. The two built the brewery until it reached a capacity of sixty thousand barrels per year.

Sprattler & Mennell

Gustav Sprattler and Christian Mennel formed Sprattler & Mennell in 1870, when they leased Christian Braun's brewery. The partnership was successful, and in 1876, they moved into a new brewery that could put out eight thousand barrels per year. They continued to build up the brewery over the years, but Gustav Sprattler died in 1885. In 1990, Sprattler's interests in the company were finally absorbed by Christian Mennel. By that same year, the brewery was producing forty thousand barrels per year.

Graham Brewery

In 1887, James A. Graham opened the Graham Brewery on Cedar Street in Paterson. Graham had worked in breweries from a young age and spent the nine years prior to 1887 as a brewery superintendent, making him one of the more experienced brewers in New Jersey. Thanks in part to its desirable location in the center of the city and near the New York, Lake Erie & Western Railroad, Graham's brewery was an immediate success.

The Katz Brothers and the Burton Brewery

The Burton Brewery began as Katz Brothers, a brewery formed by Philip and Bernard Katz in 1877 on the corner of Godwin and Bridge Streets. Its original capacity, twenty-five barrels per day, proved to be inadequate, and supply quickly fell behind demand.

Luckily for them, the Burton Brewery went up for sale around this time. The Burton Brewing Company had been formed by a group of men that, despite the wealth of its members, knew very little about the brewing industry. Internal conflict arose, and they opted to liquidate everything, which meant selling the brewery, which was one of the largest in the state. When the Katz brothers took control in 1882, it was a perfect match.

Its beers, which included **XXX** and Canada Malt Ale, became so popular that orders came in from all across the country. In 1888, Katz

Katz Brothers Brewery in Paterson, New Jersey. *Rutgers University.*

Brothers expanded, constructing a new addition that would allow it to produce lagers, whose demand was quickly growing. In twelve years, Katz Brothers had gone from 25 barrels per day to 130,000 per year.

The Paterson Consolidated Brewing Company

The four breweries came together in 1890, a year after an English syndicate offered to buy them all as well as others. That group sought to buy several breweries in Patterson to form a stock company in England. While the negotiations ultimately failed, it brought the breweries together, and they realized the advantages they would have if they united. In 1890, the four largest brewers in the city formed the Paterson Consolidated Company. The company was led by Bernard Katz (president), Philip Katz (vice-president), James Graham (second vice-president), Christian Mennell (treasurer), Louis Braun (secretary) and Christian Braun (general brewer).[49]

Hinchliffe Brewing and Malting Company | Paterson

John Hinchliffe began brewing under the banner of Hinchliffe & Co. in 1861 at an old brewery in Paterson, New Jersey. Hinchliffe later went into business with John Shaw and, later, Thomas Penrose, so by the late 1860s, the brewery was known as Shaw, Hinchliffe & Penrose. In 1872, they expanded the brewery to be able to keep up with increased demand. That expansion included the building of a brand-new, state-of-the-art malt house that became known as one of the finest in the state.

In 1878, Thomas Penrose retired, and Shaw and Hinchliffe bought his stake in the business. Just three years later, in 1881, John Shaw died, leaving Hinchliffe to run the business alone. He did so until his own death in September 1886.

Having been raised by their father and trained in the various departments throughout the brewery, Hinchliffe's three sons—John, William and James—continued the business, now under the name Hinchliffe Brewing and Malting Company, which was incorporated in May 1890. They carried on the tradition of quality beer for which Hinchliffe, Shaw and Penrose had been known. In particular, their

Original B and lager beer helped increase demand to the point where they had to build a new facility, reaching a capacity of about seventy-five thousand barrels per year in by the end of 1890. Ultimately, Prohibition caused the end of Hinchliffe.[50]

Other Breweries

In addition to the breweries already listed, there were many smaller ones that operated throughout northern New Jersey, but those previously mentioned were the major players that caused Newark, New Jersey, to become the pre-Prohibition beer capital of the United States. At its peak in about 1900, New Jersey boasted fifty-one breweries that produced 2.5 million barrels per year. While the majority of that came from the Ballantines, Kruegers, Wiedenmayers and the other large breweries, smaller breweries certainly played their parts.[51]

The Greenville Brewing Company was one of those smaller brewers. While it may not be recognized as one of the larger breweries of its time, it had a successful run nonetheless. Founded in 1890 by Charles Gerger, Adolph Becker and Daniel Kohl, Greenville Brewing produced German ales at 235 Bartholdi Avenue in the Greenville section of Jersey City. The owners sold, and the brewery became the Columbia Brewing Company on May 1, 1906. Columbia would operate until 1920, but it wouldn't survive Prohibition.[52]

Another smaller brewer that had an impressive run was the Seeber Brewery Company, also known as Rising Sun Brewing Company. Seeber Brewery opened in 1887 on Marshall Street in Patterson, New Jersey. It lasted until Prohibition and became famous throughout the 1920s for being raided more than a few times. In 1930, when federal agents from Philadelphia raided it, they were met by a dozen gunmen, who killed the agent in charge and took off.[53]

The turn of the twentieth century was met with the addition of several breweries to the North Jersey brewing landscape. In 1899, the Winter brothers—Michael, Wolfgang and Aloysius, all of whom were born in Germany—sold their Pittsburg-based brewery and moved to Orange, New Jersey, two years later. In 1902, they founded the Orange Brewing Company on Hill and Prince Streets and produced 100,000 barrels of their pilsner, porter and ale per year until Prohibition.[54] The same year the Winter brothers founded Orange Brewing, the Hudson County Consumer Brewing

Company was formed in Union City. Like many of its competitors, it did well until Prohibition.

There were, of course, many other breweries that opened and closed long before Prohibition. Here are a few of them:

- Roemmelt & Leicht Brewery | Jersey City | 1857–79
- Joseph Harth Brewery | Newark | 1860–82
- Eller & Beyer Brewery | Elizabeth | 1865–82
- John. F. Wagner Brewery | Elizabeth | 1865–84
- Morton Bros. Brewery | Newark | 1874–75
- Thomas Teneson Brewery | New Brunswick | 1874–75
- Henry F. Cox Brewery | Jersey City | 1874–77
- John Heinickel Brewery | Newark | 1874–82
- Charles Neitzer Brewery | Newark | 1874–84
- John Neu Brewery | Newark | 1874–88
- Red Star Brewery | Paterson | 1877–82
- P.J. Eckert Brewery | Elizabeth | 1878–84
- Ph. Pfannebecker Brewery | Paterson | 1878–88
- Jumbo Brewery | Newark | 1880–92
- Jacob Piez Brewery | 1884–86
- City Brewery Company | Elizabeth | 1885–1920
- Bavarian Brewery | Jersey City | 1890–95
- Home Brewing Company | Newark | 1890–1920
- Rock Spring Brewery | New Brunswick | 1898–1910
- New Brunswick Brewing Co. | New Brunswick | 1907–10[55]

CHAPTER 5
PROHIBITION

1920–1933

Mother's in the kitchen washing out the jugs,
Sister's in the pantry bottling the suds,
Father's in the cellar mixin' up the hops,
Johnny's on the front porch watch' for the cops.
—a popular Prohibition song from Collier's Weekly,
September 1, 1928[56]

THE SUNDAY LIQUOR BAN

While early in the country's history beer was mostly produced through homebrewing, the mid- to late 1800s and early 1900s saw that change, as large breweries with massive capacities—especially in northern New Jersey—replaced the need for homebrewing. Beer was now produced in large breweries and consumed not only at home but also in taverns and pubs, which had become popular destinations for social gatherings.

While Prohibition wasn't very popular, the idea of temperance was nothing new. Laws had long prohibited the sale of alcohol on Sundays, and pubs were supposed to be closed. Not surprisingly, these laws were ignored throughout the nineteenth century—pubs closed the front door but opened the side doors, and law enforcement looked the other way. In August 1903, a weekly publication from Newark noted:

Either the laws as they stand now are good or bad; if good, they ought to be enforced, and if bad, repealed…In how many cities in Northern New Jersey are the Sunday liquor laws enforced? Certainly not in Newark, nor in Jersey City, nor in Hoboken, nor in Paterson, nor in Orange…It is, in fact, a case of unauthorized and illegal Local Option, as each town has the law obeyed or disobeyed accordingly as public sentiment manifests itself.[57]

The temperance movement began to pick up pace in the early 1900s, as the religious right spoke publicly on the dangers of alcohol. In 1906, Governor Casper Stokes signed a bill that had been written by a group of clergy, rightfully becoming known as the "Bishops' Law." The new law, which went into effect on July 8, 1906, increased penalties for those found ignoring the Sunday liquor ban. Not only did it fine, jail and strip the liquor license of any bar found illegally serving alcohol, but it also went so far as to ban back rooms, making it more difficult for bars to be secretly open. Still, people drank on Sundays. And still, law enforcement looked the other way.

The fact that the law was so widely ignored didn't go unnoticed, though. In 1908, the state's governor, John Fort, initiated a study on the Bishops' Law, and the results were exactly what was expected: the law was largely ignored. The next month, Governor Fort began threatening to call in the National Guard if Atlantic City didn't start to enforce the law. He never did. But less than a decade later, on December 18, 1817, Congress passed the Eighteenth Amendment to the Constitution, also known as the Liquor Prohibition Amendment.[58]

PROHIBITION

Prohibition didn't come without a fight, though. In New Jersey, people strongly opposed Prohibition for the same reason they ignored the Sunday beer ban: they wanted their beer. Over the previous fifty years, industrialization in the beer industry had resulted in large-scale breweries, a system much more efficient than what could be produced by individual homebrewers. But this came at a cost to brewers: they had to invest massive amounts of capital in order to build these breweries. Prohibition would be devastating. They would have to shut their doors for one thing, but also, what would they do with these huge facilities—some of which comprised more than a dozen buildings encompassing entire blocks—if they couldn't make beer anymore?

A crowded bar in New York City, moments before Prohibition went into effect. *Library of Congress.*

The United States Brewers Association tried to slow down the temperance movement in 1909, when it published *A Textbook of True Temperance*. By pointing out the history of beer with the Founding Fathers, the data that showed that brewery workers were among the highest-paid workers in industry and the linkages that other industries have to brewing industry, they showed evidence of the contributions that the brewing industry made to the economy and to society.[59]

But it was all for nothing. On January 16, 1920, despite widespread criticism, Prohibition went into effect, one year after the states voted to adopt it. Despite being one of the states that initially rejected the amendment, New Jersey was forced to abide by it. It was now illegal to make, transport or sell beer over 0.5 percent alcohol by volume.

This, of course, had devastating effects for breweries across the country, not in the least those in New Jersey. Newark, the beer capital of the country, lost two thousand brewery jobs alone. When you account for the other breweries throughout the state, as well as the industries that relied on them (pubs, taverns, hotel lounges), that number increased quickly. At the time, about 15 million acres in the United States were used to grow barley, hops, rice and other ingredients used mostly for the production of beer. That land now had to find new use, as did the factories that produced barrels, bottles and caps, as well as other brewing equipment.[60]

Many breweries shut down completely, while others attempted to survive by either producing soda or low-alcohol "near beer" (which was officially labeled as cereal beverage, not beer) or illegally producing full-strength beer. Ballantine's owners saw Prohibition coming and were able to cope with the new times by expanding into other products like malt syrup while also diversifying into insurance and real estate. John H. Ballantine, Peter Ballantine's son, took control of the brewery in 1883. He soon got into an argument with his oldest son, John, who left the family business and opened the Neptune Meter Company. When Prohibition came, many family members left Ballantine and joined John at Neptune.[61]

Other breweries tried to make an honest living by selling near beer and soda. One of those breweries was Krueger's, and this proved very beneficial for it. When Prohibition ended, Krueger's was the only brewery in the state that was ready to brew right away. It was so overwhelmed that beer was sold right outside the brewery, and it took two days for order to be restored once it opened back up.

Christian Feigenspan, the "Pride of Newark" brewer, fought the Eighteenth Amendment as hard as he could. Serving as the president of the United States Brewers Association, he led a lawsuit that argued that Prohibition was illegal and that New Jersey didn't have to obey it because it was one of the few states that had refused to ratify it. Needless to say, he lost the case. Years later, in 1927, he petitioned to be able to give beer that had been sitting in tanks since before Prohibition to his shareholders. His petition was denied, and he was forced to let 300,000 gallons of beer go into the Passaic River.

Other breweries changed businesses completely, opting not to produce near beer (which no one wanted) or soda (which didn't see an increase in demand once beer was illegal). Wiedenmayer of Newark converted and began producing ice cream and other dairy products. Lembeck & Betz Eagle closed and was later sold and converted into a refrigeration plant.

Finally, there was the beer mob, something known well by many New Jersey residents. It's rumored that during Prohibition, Hensler Brewery was controlled by mobster Waxey Gordon. Waxey also bought the old Sprattler & Mennel Brewery, which had closed at the onset of Prohibition. There, he continued to brew full-strength beer, which he smuggled out through fire hoses that ran in sewer pipes. There was also the Hauck Brewery, which was run by Max Greenberg and Max Hassel, who operated it under the Harrison Beverage Company name. Of course, there was also Breidt Brewery, which was involved in the 1923 arrests of eleven men from three breweries (the others were Rising Sun and Hygeia) who had been selling full-strength beer, thanks in part to the help of a bribed Prohibition agent and four state officials. Despite all the evidence against the eleven men, the case was dismissed in August 1926, when Saul Grill, the undercover agent who had witnessed the bribes, disappeared.[62]

LIKE THE SUNDAY LIQUOR ban, Prohibition wasn't taken seriously by many, especially in New Jersey. Being New York City's neighbor, New Jersey was populated with a large number of immigrants from Germany, Ireland and Poland. In their native lands, alcohol was a part of life. Beer was often

New Jersey women picketing at a Prohibition hearing at the capitol, April 1926. *Library of Congress.*

drank with, or even as, breakfast, and social drinking was a part of everyday life. When Prohibition came along, they thought it was a joke. Just like the Sunday liquor ban, the everyday beer ban was never going to succeed in Jersey. Even the state's governor, Edward Daniels, publicly announced that he was "as wet as the Atlantic Ocean." His anti-Prohibition platform helped him get elected as both governor and senator.[63]

There had been a push in the early 1920s to legalize beer for medicinal purposes, as alcohol had previously been used for its calming and sedative effects. Shortly before retiring, General A. Mitchell Palmer, the man previously known for launching the "Palmer Raids" during the Red Scare, issued a ruling stating that the Volstead Act did not prohibit beer from being used for medicinal purposes, thereby allowing doctors to prescribe beer to patients. And that's exactly what they did. Breweries throughout the country, including in New Jersey, filed for permits to produce medicinal beer, as doctors prescribed beer to patients for a number of health issues. One Chicago doctor reportedly prescribed beer to seven thousand patients over just a matter of weeks.[64]

It didn't last long, though, as Congress quickly fixed the problem. In 1921, Congress passed the Willis-Campbell Act, written by Senator Frank Willis and Representative Robert Campbell, which prohibited the use of medicinal beer.

To ENFORCE THE BAN on alcohol, the Prohibition Bureau set up regional offices throughout the country. In 1926, fifty-four-year-old Ira L. Reeves, an army veteran who had received a citation for bravery and became a colonel in World War I, took over the New Jersey office, located in none other than Newark. It was fitting that it would be Newark, as the city had become the bootlegging capital of the country, providing an estimated 40 percent of all illegal alcohol in the country during Prohibition.[65]

Reeves proved to be a good at the job. As soon as Prohibition began, men saw an opportunity to profit by bootlegging and selling beer illegally. In his book, Reeves wrote that he thought of himself as the Prohibition St. Patrick of New Jersey, hoping to "be able to make bootleggers and their kind as scarce as snakes in Ireland." He and his agents quickly went to work, conducting raids throughout the state at speakeasies, breweries and anywhere else there was illegal beer flowing. Despite his attempts, Reeves was widely unsuccessful, and Prohibition was ignored not only throughout the state but also in Newark, right under his nose.

One reason was the fact that even law enforcement officials didn't support Prohibition, and many refused to enforce it. They were easily

New York City deputy police commissioner John A. Leach (right) watching agents pour liquor into the sewer following a raid during the height of Prohibition. *Library of Congress.*

bribed to look the other way or, in some cases, were so sympathetic that they didn't even require a bribe. For example, in 1925, a former Boston Celtics basketball player named Moe Katzman opened a restaurant called the Mansion House directly across the street from the Hackensack Courthouse. He was so well liked that he didn't have to pay off local officials—he was just protected. He was arrested one day by an undercover officer who was unaware of Moe's position as a protected man. When Judge Charles McCarthy saw him at the courthouse, he reportedly said, "Get out of here, Moe." And that was that.[66]

At times, law enforcement wasn't merely unhelpful but actually got in the way. One example was the night of January 20, 1927, when Reeves sent three of his agents to raid a warehouse on Broad Street in Trenton. The agents arrived to find that a mob had formed in front of the warehouse, threatening them if they tried to raid the warehouse. One of the agents

A woman pours liquor from a cane into her cup at a soda fountain during Prohibition. *Library of Congress.*

fired a warning shot into the air, catching the attention of a nearby patrol cop. How did the local cop diffuse the situation? He arrested the agents for carrying guns without a license. Reeves quit four months later, after serving just eight months, and went on to write a book, *Ol' Rum River: Revelations of a Prohibition Administrator*, in which he criticized the Prohibition movement:

After months of the most strenuous effort—doing my office duties during the day, raiding at night, often working eighteen hours a day for two or three weeks at a time—I took inventory of what had been accomplished. I had raised the price of alcoholic beverages and reduced the quality. *That's all. There were just as many bootleggers, now making bigger profits than before. There were doubtless just as many wildcat stills, cutting plants, breweries, ale plants, roadhouses, saloons, and speakeasies as before my ambitious crusade. True, they were running more risk, but they were getting better prices, which was a real measure of compensation for the added hazard. I then realized what all the other administrators in the United States had learned—the prohibition laws are unenforceable.*[67]

While Reeves never became a drinker, he did become an anti-Prohibitionist and went on to make a considerable living speaking out against the Eighteenth Amendment. And he wasn't alone. This sentiment, it turns out, was shared by many, as the country began to realize that the Prohibition experiment had failed. As Reeves argued, it turned good people into criminals. The first step to repealing Prohibition was taken on March 22, 1933, when President Franklin Roosevelt signed the Cullen-Harrison Act, an amendment to the Volstead Act, which allowed beer up to 3.2 percent ABV to be made and sold. Later that year, on December 5, Prohibition was repealed.[68]

The repeal went into effect at midnight on April 7, 1934, and it was celebrated nationwide. According to sources, more than 1 million barrels (more than 31 million gallons) of beer were sold that first day. In New Jersey, Krueger Brewery had been the only brewery that was ready to begin brewing right after Prohibition ended (as many others had converted to produce soda, ice cream or other goods or had closed outright). Krueger opened on April 7 and began selling beer from the brewery straight into cups. According to some newspaper reports, it took two days for the line in front of the brewery to return to order.[69]

CHAPTER 6

PROHIBITION AS THE CATALYST FOR ORGANIZED CRIME

Prohibition was widely unpopular, especially in New Jersey. It was, after all, pushed through by the religious right, most of whom lived in the Midwest. New Jersey, on the other hand, was a diverse mixture of immigrants, many of whom came from Ireland, Poland, Germany and other European countries where drinking was commonplace, even for young children. When Prohibition went into effect, these populations thought it was a joke and looked to continue their drinking regardless.

The economic effects of Prohibition were as disastrous as the social ones. Before January 1920, 1,150 breweries were producing 50 million barrels of beer per year. The industry provided more than $150 million per year in taxes, employed tens of thousands of people and supported dozens of industries that, in some part, relied on beer.

Brewery owners had a choice: they could keep brewing beer illegally; they could produce low-alcohol "near beer" even though there was little to no demand for that; they could convert their factories and produce other goods, like soda or dairy products; or they could simply walk away. Most chose the latter. They had already made their fortunes, and many were thought to be exemplary citizens and often served in public office. It simply wasn't worth the risk to illegally produce beer, and they didn't want to bother producing near beer, soda or ice cream. Most just called it quits; of the 1,150 pre-Prohibition breweries, 986 went out of business.

Charles "Lucky" Luciano's 1931 mug shot. *New York Police Department.*

But brewers calling it quits didn't make the people want beer any less. There was still demand for it, despite the low supply. Applying simple microeconomics tells us one thing: if you were willing risk it, there was money to be made. And there were plenty of people willing to take that risk. Even better, there were breweries available for purchase, and they were cheap. After all, who else wanted to buy an old brewery during Prohibition?[70]

High demand, low supply and cheap equipment and ingredients formed the perfect environment for crime to thrive. And during Prohibition, that's exactly what happened, especially in New Jersey; it's estimated that during Prohibition, about 40 percent of all smuggled alcohol in the United States came through Newark. As Mike Dash wrote in *The First Family: Terror, Extortion, Revenge, Murder, and the Birth of the American Mafia,* "Nothing like it had ever happened before. An entire industry—one of the most important in the country—had been gifted by the government to gangsters."[71]

Prohibition gave way to some of the most famous gangsters of all time: Al Capone, Abner "Longy" Zwillman, Charles "Lucky" Luciano, Meyer Lansky, Frank Costello, Ben "Bugsy" Siegel, Dutch Schultz, Lepke Buchalter, Max Hassel, Waxey Gordon, Max Greenberg, Ruggiero "Richie the Boot" Boiardo and others. Most started off simple: they owned small businesses, ran protection schemes or ran gambling rings. Many got their start in alcohol by smuggling whisky, usually from Canada.

Beer, it would turn out, was just the next logical step. They already had distribution networks set up with local restaurants and

Meyer Lansky. *Library of Congress.*

speakeasies that they could utilize to sell their beer. Breweries, of course, were cheap to purchase since no one else wanted them. They would get licensed (not using their real names, of course) to produce near beer, which was made by producing full-strength beer and then removing the alcohol. They would simply sell the beer before the alcohol was taken out. Often, they would run underground pipes to transport the beer to neighboring buildings. When inspectors came in for a visit, they'd only find the little amount of near beer that had been produced to keep up the façade.

Longy

The biggest bootlegger on the East Coast was Abner Zwillman, commonly known as "Longy." Like many other gangsters of his time, he came from humble beginnings. Born in Newark in 1904, he grew up as one of seven children of poor Russian immigrant parents. He began working at a young age to earn money for his family after his father died. He started selling fruit on the street when he was just twelve years old and later worked as a milk delivery boy. He was well liked in his neighborhood. Although he was known as being quiet and polite, he wasn't afraid to fight. Occasionally, Irish kids would come into his neighborhood, the Third Ward, and yell, "Reef der langer!" which means, "Get the tall one!" in Yiddish. As "Der Langer" grew up and associated with different group of people, his nickname was shortened to "Longy."[72]

Growing up in Newark's Third Ward, he began working for Joseph Mann, the local political boss. It was here where Longy began to learn the business of the streets. He came to realize that there were two ways to make money in America: politics and gambling. He had developed a network of local business owners and residents as a milk delivery boy as a young teenager, so he began putting it to use when, at just sixteen, he began running a numbers lottery with Manny Kimmel, his boyhood friend. Gamblers would bet a three-digit number, and every day, one of those numbers would win. The winner won six hundred times his bet, and Zwillman and Kimmel earned $400 for every $1,000 that was gambled.

Getting involved in illegal business had its risks, of course, including the risk of treading on another gangster's turf. One of the men was Leo Kaplus, who threatened to kick Zwillman in his testicles to teach him a lesson. Zwillman responded by tracking Kaplus down and shooting him in the same spot where he had threatened him. With Kaplus out of the way, Longy's numbers game was able to grow quite large, and he made a hefty profit from it.[73]

But money wasn't the most important thing that came from his lottery—it was the network of people, businesses and runners that he developed. So, when the Volstead Act went into effect, enforcing Prohibition, Longy was ready to seize what he knew to be a huge opportunity. He saw the demand for alcohol, and he knew that Prohibition would be difficult, if not impossible, to enforce. As he once said, "You got to be crazy not to try and make money selling something everybody wants."[74]

One of Longy's connections was Joseph Reinfeld, owner of a local tavern, who had a relationship with Seagram's Distillery in Canada. While

Above: A Philadelphia Phillies ad featuring Ballantine. *Hugginsandscott. com.*

Left: Dogfish Head's Palo Santo Marron is infused with coconut using Dogfish Head's "Randall." *Chris Morris.*

Whole hop cones are added during the boiling part of the brewing process. *Chris Morris*.

A can of Ballantine Beer, brewed in Newark, New Jersey. *Art LaComb*.

Krueger's Cream Ale was a favorite among New Jersey beer drinkers. *Art LaComb*.

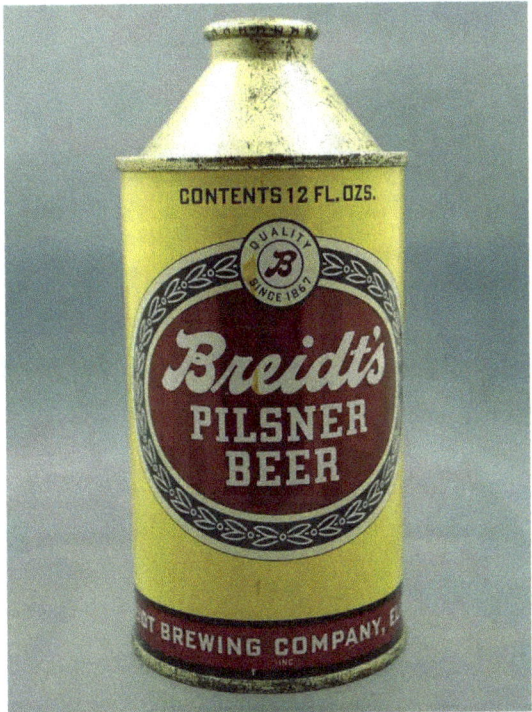

A can of Breidt's Beer from Elizabeth, New Jersey. *Art LaComb*.

Above: Triumph's fermentation tanks sit behind the bar. *Chris Morris.*

Left: Two beers from Hillsborough's Flounder Brewing. *Flounder Brewing.*

Right: Four taps pour in Flounder Brewing's tasting room. *Flounder Brewing.*

Below: Several dozen barrels line the wall at River Horse's Ewing brewery. *Chris Morris.*

Artwork lines the walls around the River Horse brewery. *Chris Morris.*

In the River Horse tasting room, fresh grains and hops are on display to allow visitors to smell the ingredients. *Chris Morris.*

Bolero Snort's season "Grazer," a dry-hopped American wheat. *Chris Morris*.

Fermentation tanks sit at the center of the J.J. Bitting's dining area. *Chris Morris*.

J.J. Bitting's brewer, Mehmet Kadiev, tests a beer from one the brewery's fermentation tanks. *Chris Morris.*

World of Beer in downtown New Brunswick boasts fifty rotating beers on tap alongside more than five hundred bottles. *Chris Morris.*

World of Beer in New Brunswick. *Chris Morris.*

Five beers from North Bergen's New Jersey Beer Company. *Andrew Barrack/Great Heights Photography.*

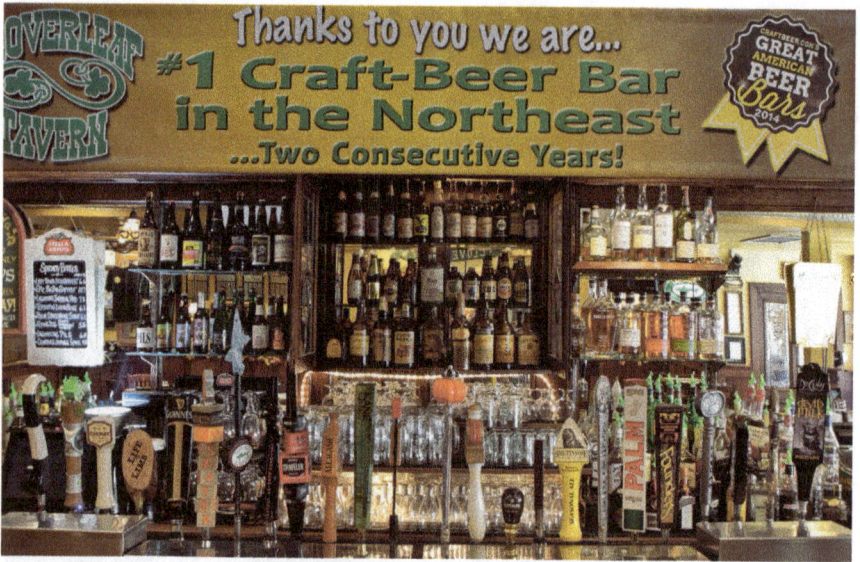

Caldwell's Cloverleaf Tavern has been named the best craft beer bar in the Northeast. *Cloverleaf Tavern.*

Ron Witkowski, Mark Spezio and the staff of the Love2Brew homebrew store in North Brunswick. *Courtesy of Love2Brew.*

Right: BOAKS' Double BW Belgian Wheat. *Peter Culos*.

Below: A beer sampler from Princeton's Triumph. *Interchangeableparts*.

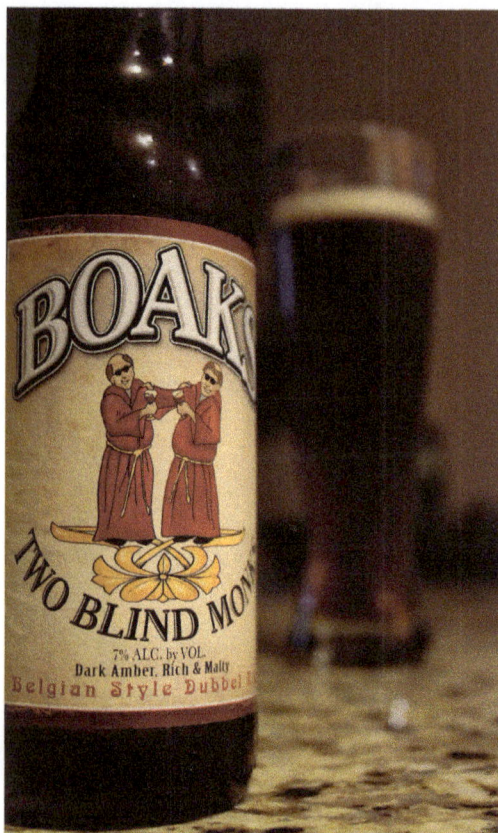

Left: BOAKS' Two Blind Monks Belgian Dubbel Ale. *Chris Morris*.

Below: A frothy head on Cricket Hill's Hopnotic comes from the prickly carbonation. *Chris Morris*.

River Horse's hop garden, recently planted outside the Ewing brewery. It hopes to be brewing a fresh harvest ale in the next few years. *Chris Morris*.

A beer sampler showcases some of Long Valley Brew Pub's fine craft beers. *Robert Greco*.

Above: Krogh's sampler. *Julian Huarte.*

Left: Krogh's oatmeal stout. *Julian Huarte.*

Members of New Jersey Craft Beer, Carton Brewing, New Jersey Beer Company and others at Atlantic City Beer and Music Festival. *Jon Miller/New Jersey Craft Beer.*

Bob Olson and Andrew Maiorana of Bolero Snort after winning the People's Choice award at Atlantic City Beer and Music Festival. *Jon Miller/New Jersey Craft Beer.*

Left: Mike Kivowitz holds up 2015 New Jersey Craft Beer membership cards, surrounded by other NJCB representatives. *Jon Miller/New Jersey Craft Beer.*

Below: Moshe Atzbi runs the Beer Senate at Hailey's Harp & Pub in Metuchen. The theme of this senate is "Christmas versus Hanukah." *Chris Morris.*

Reinfeld may have preferred to go out on his own, Longy's offer of a fifty-fifty partnership was hard to turn down, especially since going out on his own would mean competing against Longy, which he didn't want to do. Together, the two of them imported what the IRS estimates to be about 40 percent of all alcohol smuggled from Canada. Longy and Reinfeld's operation was extremely successful—while other bootleggers were selling homemade product, they were selling quality, brand-name whisky.[75]

With his distribution network in place and breweries cheap to purchase, it only makes sense that Longy would get involved in the beer business. He did, and throughout Prohibition, Zwillman made millions.

Clearly, Zwillman was just as smart as everyone gave him credit for early in his life. He'd prove them right again. As the years passed, Longy knew that Prohibition—which earned him his millions—wouldn't last. He knew that he had to plan for a future without it. His first step was to organize what became known as the Big Six Mafia Ruling Commission, which would, in a way, regulate gambling and bootlegging operations around the New York City area. The other members were Charles "Lucky" Luciano, Meyer Lansky, Frank Costello, Joe Adonis and Ben "Bugsy" Siegel.

The Big Six proved to be a powerful and profitable group for all the members involved. Longy convinced the group to invest in a number of restaurants throughout the area that would sell Longy's beer. Not only did it increase Longy's retail presence, but it also helped increase his ties with his New York partners, who were in charge of the restaurants' gambling operations.[76]

The group also caused some problems, especially when it came to non-members like Dutch Schultz who were upset about being left out. Schultz was born in Manhattan in 1902 as Arthur Flegenheimer, the son of two German Jewish parents. He grew up an avid reader and was a fan of Horatio Alger, a nineteenth-century writer who wrote stories of young men overcoming poverty. Schultz was inspired and worked hard to support his family through work as a roofer, newspaper delivery boy and a printer's assistant. But he soon realized that there was money to be made in gambling and crime. He was arrested as a member of the Bergen Gang and served fifteen months. While he spent much of the 1920s as a low-level criminal, things changed in 1928.

In that year, Schultz teamed up with Joseph Noe (also known as Joey Noey) and opened the Hub Social Club, a speakeasy in the Bronx, before investing in other bars throughout the borough. They later expanded into bootlegging by transporting beer from Jersey City not only to their bars but also to others that they could threaten into buying their product. Over time, Schultz and

Noe grew their operations throughout New York City through their gang, alongside Bo Weinberg, Fatty Walsh and Vincent Coll—the latter became an enemy of Schultz and was eventually killed by him. They were able to grow through a reputation for violence that gave their threats credibility. On occasion, this violence caused issues with rival gangs. They kidnapped and beat John and Joe Rock, two brothers who ran a rival bootlegging operation in the Bronx. On October 15, 1929, Noe was killed in a shootout with Louis Weinberg, a member of rival Jack "Legs" Diamond's gang.[77]

Known for his violence, Schultz made a lot of noise with very public shootings, and the Big Six didn't like that. They decided that he was going too far when they heard his plans to kill Thomas Dewey, a New York prosecutor, in what would be a very public murder. So, they agreed to take Schultz out instead, hiring Lepke Buchalter's Murder Inc. to do the dirty work, which it did on October 23, 1935, in the bathroom of Dutch's favorite restaurant, the Palace Chop House. At least that's the leading theory on Dutch's death, which has never truly been solved. Schultz survived the initial gunshot and was taken to Newark City Hospital. There, his wound became infected, and Schultz developed a fever of over 105 degrees. His illness caused him to rant in delirium, so police had a stenographer sit by his bedside and take notes of whatever he said, thinking that he might reveal clues into who had shot him. He didn't, and on October 24, 1935, the day after being shot, Dutch Schultz died.[78]

Around the same time that Longy was building his beer-running empire, another Newark resident, Ruggiero Boiardo, was making similar strides. Settling in Newark when he was twenty, Ruggiero, like Longy, started out delivering milk, eventually transitioning to beer and whiskey. He did his business out of a phone booth, earning the nickname "Richie the Booth." With a strong Newark accent, this was transformed to "Richie the Boot."

Richie's tough attitude and good business sense helped him control Newark's First Ward. But when he started to expand out of the First Ward and into the Third, there were problems, as the Third Ward was Longy's territory. Richie's gang began demanding that Third Ward restaurants and saloons buy their product from Richie and shot up bars that refused; it is said that they killed one of Longy's men. The gang war, also called the Newark Beer War, lasted a few weeks, reaching a peak when Longy had his men gun down Richie on Newark's Broad Street. The Boot was shot eight times but survived. Longy, being as smart as he was, decided that he wasn't going to sit back and wait for Richie to retaliate. Instead, he had his men find Richie's guys and break their knees.[79]

Louis "Lepke" Buchalter entering a courthouse, handcuffed to J. Edgar Hoover (at left). *Library of Congress.*

Richie attempted to retaliate while Longy was staying at the Riviera Hotel, but again, Longy was too smart and stayed out of sight. To get inside, Richie sent two women to the hotel. When they asked to see Longy, the front desk called up to his room and warned him that something seemed strange. Longy allowed the two women to come up. When the elevator door opened, two bodyguards met them with guns pulled. The guards searched them and, not surprisingly, found guns hidden under their dresses. What was surprising, though, was that they weren't women at all, but two eighteen-year-old boys, who were now shaking and sweating in fear. After an interrogation, Longy let the two boys go, with a note to have Richie call him to work out a truce. Eventually, an agreement was reached, with both gangsters respecting their original territories.[80]

Longy saw the end of Prohibition coming and prepared accordingly. While he continued his illegal bootlegging operations throughout the early 1930s, he also expanded to more legitimate businesses both before and after the repeal of Prohibition: he bought stakes in two steel companies, the Hudson & Manhattan Railroad (now owned by the Port Authority and known as the PATH), a Newark truck dealership, nightclubs, liquor stores and parking garages and even invested in Bugsy Siegel's Flamingo Casino in Las Vegas. Longy also invested in a few small production companies, perhaps due to his romantic relationship with actress Jean Harlow. As the story goes, Longy even bribed Harry Cohn of Columbia Pictures with $500,000 to get her a two-movie deal.[81]

Like many of the gangsters who weren't killed in the 1920s or 1930s, Longy's downfall came largely at the hands of the Internal Revenue Service (IRS). His problems with authorities perhaps began during Prohibition, when one of Joseph Kennedy's shipments of Haig & Haig whiskey was hijacked outside Brockton, Massachusetts. Although Longy denied it until the day he died, Kennedy accused Longy, his rival, of being responsible, claiming that he was the only person capable of pulling it off. Longy blamed much of his later legal troubles on this old grudge.

Kennedy, of course, has several sons who pursued political careers, including Robert, who worked as an aide during the Kefauver hearings. The hearings, run by Senator Estes Kefauver, were done by the U.S. Senate Special Committee to Investigate Organized Crime in Interstate Commerce, and on March 27, 1951, Longy testified. Televised throughout the country, his testimony helped him gain national fame. Later that year, in August and September, *Collier's Weekly* published a full exposé that backed Kefauver's claim that Longy was the biggest gangster in the country.[82]

In addition to identifying him as the country's leading gangster, the committee's final report also called for further investigation into Longy, which the IRS began in June 1952. Also under investigation was one of Longy's partners, Joe Adonis, who went on trial in 1954 for racketeering. Deciding that he was too good-looking for prison, Adonis opted for deportation to Italy. His absence left Longy short one powerful ally. Longy would lose another in 1957. Stemming from an old dispute with Vito Genovese, one of Lucky Luciano's men, Frank Costello was shot in his apartment on May 2, 1957. Although he survived the hit, Costello wouldn't go back to work; instead, he opted for early retirement.[83]

Even before Costello's retirement, Longy found himself in plenty of legal trouble. On January 9, 1956, Longy entered trial in Newark's federal court. The key piece of evidence that prosecutors used was actually something that Longy himself had given them. Years earlier, in 1947, Longy visited Newark's IRS office with his lawyers and accountants to discuss his tax situation. There, he was asked to provide a net worth for himself and his wife, Mary, which he provided to be $623,672.60.

Despite this evidence, the jury was unable to reach a verdict after thirty hours of deliberation, and Longy was released. But his freedom was short-lived. The IRS and Federal Bureau of Investigation (FBI) had been prepared for the possibility that some jurors would be bribed—and they had been. All the jurors who had taken the bribes were arrested, as were the men who had offered the bribes. One of those men was Sam "Big Sue" Katz, a known aide to Longy.

Longy, ever a smart man, knew that he was in trouble and became noticeably depressed and withdrawn. He was likely to be tried for tax evasion but now faced charges for the bribing scheme and massive legal fees and fines for his tax penalties. He was also developing health issues.

On February 25, 1959, after he indulged in a night of heavy drinking, Longy's wife, Mary, woke up at about 2:00 a.m. and found him pacing the room. He gave her a sleeping pill and told her to go back to sleep. The next day, after waking up along and assuming that Longy had gone to work early, Mary went into the basement and found her husband dead, hanging from a ceiling rafter with tranquilizers in his pocket and a half-empty bottle of bourbon nearby.[84]

The Jersey Trio

The Jersey Trio was a group of gangsters that ran much of the northern New Jersey beer scene from the late 1920s through 1933, a year that proved to be important in the story of Jersey gangsters, and not just because it saw the end of Prohibition. The group was made up of Irving Wexler (known by his nickname "Waxey Gordon" or just "Waxey"), Mendel Gassel (better known by his Americanized name, Max Hassel) and Max Greenberg (known widely as "Big Maxie"). Together, they ran an estimated sixteen or seventeen breweries and made millions.

THE FIRST OF THE Jersey Trio was Max Hassel, who traveled an uncommon path into bootlegging. While many of the gangsters of the time were already heavily involved in crime prior to Prohibition, Hassel wasn't. Hassel was born to Jewish parents in Latvia on April 24, 1900, as Mendel Gassel and was the third of five children (Fannie and Calvin were his older siblings, and Morris and Lena were younger). His father, Elias, decided to move his family to America in about 1908, as Tsar Nicolas II ruled a heavily anti-Semitic Russia. Elias made the journey with Fannie first, arriving at Ellis Island before settling in Reading, Pennsylvania. Mendel, along with his mother, Sarah, and his three remaining siblings, made the eleven-day journey across the Atlantic aboard the SS *Kursk* three years later, arriving at Ellis Island on September 22, 1911, before meeting his father and sister in Reading. (Meyer Lansky, who later became another powerful gangster, also came to America via the SS *Kursk*.)

Upon landing in America at eleven years of age, Mendel immediately changed his named to Max Hassel, although not legally. His father also followed suit and began going by the name Ellis Hassel. While all evidence points to Max being a good, well-mannered student during his first few years in the country, he dropped out of school when he was fourteen—not an uncommon act at the time, as teenagers were eager to begin work, and parents were equally as eager to have more income.

Max spent his first few working years at department stores before selling newspapers for the *Reading Telegram*. He was known by his customers as polite and smart, always quick to give the proper amount of change. While his father earned a comfortable enough living as a tailor, Max was apparently uninterested in following his father's path. At sixteen years old, Max launched his first business, Berks Cigar Company, with his closest friend, Israel "Izzy" Liever. The town of Reading, which sat in Berks County, had long been a popular cigar producer, with thousands of area residents employed in the industry. The two did well in the business, earning a reputation for producing some of the finest cigars around. In 1919, they opened a second company, Universal Cigar Stores, a small chain of retail stores. The stores proved a valuable opportunity to learn about the oncoming Prohibition, something often spoken of by businessmen waiting in line. Some of them even talked about the profit that could be made from this "Great Experiment." As Hassel biographer Ed Taggert noted, "Max listened more than he talked…By the time he was 19, Max had 20/20 vision regarding Prohibition. There would always be a market for booze and he jumped at the chance to strike it rich. Many bootleggers would turn out to be little more than street fighters, but from the beginning, Max took aim at the heavyweight crown."[85]

Max's first involvement in Prohibition alcohol was through his business called Schuylkill Extract Company. At the time, the government issued permits to allow for alcohol to be used by some manufacturers and for religious purposes; Max's business claimed that it would produce food extracts. Max found that it was easy to work with agents of the Prohibition Agency, as most were more concerned with the extra income than they were with enforcing the rules. Max ran into problems when the Philadelphia Prohibition office sent its own agents to investigate Schuylkill Extract. They found records of a shipment of twenty-one thousand gallons of Horke wine that had gone to M. Weiner. When they went to Weiner's store, he was said to have been missing for two weeks and had no equipment to bottle wine, and the landlord claimed that he had never seen any large shipments arrive. Hassel, at the urging of his attorney, forfeited his permit. By 1922, though, Max was back at it with a new operation under a fake name, Stanley Miller—when the enterprise was found out, Max admitted that the name had been forged. Still, Max was proving to be a smart businessman and a good liar under interrogation.

Max's first real step into the bootlegging world came when he purchased the Fisher Brewery from Harry Fisher in 1921. The problem was that he wanted the brewery to come with a brewery license, which Fisher hadn't renewed in four years. Still, the deal was made, and Max purchased the brewery under the company name Brazilian Aramzem. In April 1922, Fisher argued in front of Judge Gustav Endlich that he wanted to renew his license because he thought there was a market for near beer. Without mentioning Max or Brazilian Aramzem, the permit was renewed. As Taggert noted, Fisher was a great business for Max in his first years as a bootlegger. While a half-barrel of beer cost about $2.50 to make, it could be sold to wholesalers for $8.00 to $10.00 or directly to speakeasies for $11.00 to $16.00. Clearly, there was money to be made, and Max was just the guy to make it. Max soon expanded through purchases of Lauer Brewery and Reading Brewery. Like his cigars, Max has earned a reputation for producing the best beer.

The feds would eventually catch up to Max, and after a long battle, his Reading-area breweries would be shut down. The mid-1920s saw Max

Hassel fight the IRS. While Max claimed to have made most of his money throughout the early part of the decade ($6,000 in 1920, up to $21,534 in 1924), tax expert Joseph E. Kelley disagreed, listing those numbers at about $15,375 in 1920, $113,003 in 1923 and as high as $1.08 million in 1924 (and more than $1 million again in 1925). In October 1926, the government filed a $1.24 million tax lien against Hassel. After years of fighting, Max decided that it was time to settle and move on in February 1929. On February 5, 1929, federal judge J. Whitaker approved an agreement of $150,000 and no jail time—not bad considering the original claim made by the IRS.

Another legal worry was still on Max's plate, though. Waiting outside the courtroom was Philadelphia County detective Louis Raphael, who was there to serve Max with a subpoena to appear before a grand jury in Philadelphia. The problem: Max walked out of the courtroom with his brother, Morris, and Raphael couldn't tell the two apart. As it turns out, neither could the Prohibition agent who was there with him, as he pointed to Morris when asked which one was Max. When Max and Morris parted ways, the two men followed Morris and missed the chance to serve Max. Max never did get served, and of the eighty-two suspects subpoenaed, he was the only one not indicted.

Following the ordeal with his Reading operations, Max decided to focus on New Jersey, where the climate was friendlier to bootleggers, since Jersey was overwhelmingly a wet state. His big move into the Garden State came in 1928 when he purchased the Camden County Cereal Beverage Company across the Delaware River from Philadelphia. South Jersey had high demand for illegal booze, and Max thought that he'd be just the guy to fill it. To get as much protection as possible, he hired Yates Fetterman, a recently resigned deputy Prohibition officer. Yates worked his way up to deputy of the Pittsburgh office and was transferred to Philadelphia in March 1927 to serve as that office's deputy. Maybe he was upset about not getting promoted, or maybe he didn't earn enough money. Either way, he resigned a year after his transfer and joined Max. With his knowledge of the Jersey bootlegging landscape and his connections around South Jersey, Yates became a valuable member of Max's team, helping Hassel's beer saturate the market.

Around this time, Max was forming a rivalry with Mickey "the Muscle" Duffy, who ran a large operation in Philadelphia. Mac and Mickey had two very different ways of doing business. While Max was extremely anti-violence, refusing to even carry a weapon for much of his life, Mickey was the exact opposite—he was a violent, strong-armed thug. And unfortunately for Max, he wanted Camden County Cereal Beverage.

In early 1929, Mickey and some of his men traveled to Reading to speak with Max. They drove to Isaac Marks's Colonial Cigar Store and asked Ed Marks, who was working, where to they could find Max. Ed refused to tell them, even when one of Mickey's men pulled out a handgun. When Mickey finally did catch up with Max, his offer of a partnership was refused.

Mickey didn't like to be rejected. One night, while Max was working alone in the Camden brewery, Mickey and a dozen or so of his men stormed the brewery. They picked Max up, dragged him out and dropped him on the street. The brewery now belonged to Mickey. Max, though, didn't retaliate—that wasn't in his nature. He waited a few weeks and then made a deal with Mickey, where the two would partner and share the profits. This proved a smart move, as Mickey was murdered two years later.

With his tax worries behind him for now, Max decided that North Jersey might be the best place for him to expand. That's where he met Waxey Gordon.[86]

While Max Hassel had an unusual path into crime, his future partner, Waxey Gordon, saw a more natural progression. Waxey, born Irving Wexler on January 19, 1888, earned his name by being a waxy-smooth pickpocket on the Lower East side of New York City. He worked for a while as a collector for local bookies, earning a reputation for being rough and violent.[87]

With the backing of Arnold Rothstein, Waxey continued his natural progression into bootlegging by running whiskey through North Jersey. Arnold Rothstein was born in 1882 to Abraham Rothstein, a successful and widely admired businessman, and his wife, Esther. He and his four siblings grew up on the wealthy Upper West Side, far away from the crime that took place in Lower Manhattan. Despite his upbringing, which saw his siblings do quite well in school, Arnold left home in his early twenties to become a gambler, which turned him from a quiet boy to an extrovert of a man.

Over time, Rothstein began to invest in gambling businesses, in addition to the gambling he did himself, which saw a lot of success due to his quick math skills. His reputation in the gambling world really took off when he won a game of pool against a Philadelphia champion that took thirty-two hours, from 8:00 p.m. one Thursday to 4:00 a.m. that Sunday. His fame also grew due to his role in the Black Sox Scandal of the 1919 World Series, in which several members of the Chicago White Sox threw the game, allowing gamblers who bet the Cincinnati Reds, who were huge underdogs, to win big. He's also credited with being one of the first to see the business opportunity that came with Prohibition, and he capitalized on it. With plenty of money on hand, he came up with—and bankrolled—early schemes to import liquor into the country, realizing that even with the onset of Prohibition, the demand for alcohol wasn't going to go away. Throughout the 1920s, Rothstein earned the nickname "The Brain," as he financed operations for gambling rings and bootlegging operations throughout New York and New Jersey. Two of the men he partnered with early on were Waxey Gordon and Max Greenberg.[88]

Max Greenberg, known commonly as "Big Maxie," was born in 1883 and began his career in Detroit, overseeing an operation in which "religious" people purchased wine through rabbis, which was then allowed under the Volstead Act. He then began selling this sacramental wine to bootleggers in the St. Louis area, where he began to make a name for himself.[89]

He entered the Jersey/New York scene in 1920, when he became one of Rothstein's first partners. That fall, Waxey Gordon introduced Rothstein to Greenberg, who, at a bench in Central Park, asked Rothstein for $175,000 to expand his smuggling business by buying speed boats to import Canadian booze from Ontario to Detroit. Rothstein agreed to the plan, but with a few conditions. First, he wanted to booze to be sourced from his contacts in Europe, not Canada. Second, he wanted the booze distributed through New York, not Detroit. And third, Greenberg had to take out the biggest life insurance policy he could get with Rothstein's insurance company, and he had to make Rothstein the beneficiary. Greenberg agreed, and the two were in business. Rothstein arranged a Norwegian ship to transport the booze across the Atlantic. Speedboats then carried the shipments to Long Island and the Jersey Shore, where trucks picked them up and took them to a warehouse. It was all made possible thanks to the easily bribed Coast Guard and local police.[90]

Rothstein also backed a number of other bootleggers throughout the area, including the recognizable names of Bugsy Siegel and Meyer Lansky. For a year, these Rothstein-backed enterprises returned huge profits. But eventually, they started running into problems.

The first came at the greedy hands of Waxey Gordon, who was splitting time between his Philadelphia- and New York–area markets. While Rothstein took pride in delivering quality products, Waxey was just like every other gangster: greedy. He began cutting the quality product with cheap hooch, which caused problems between him and Rothstein.

Rothstein's problems grew worse as internal feuds began to grow in what became known as the "War of the Jews." Waxey had been dealing with Joe "The Boss" Masseria, who hated Jews but needed Waxey's help to deliver scotch to his customers. Meyer Lanksy, also working with Rothstein, found out about the deal and had his gang ambush the transport outside

Atlantic City. Lansky was seeking revenge on Masseria for an older grudge stemming from his Brooklyn gambling business, and he had never liked Waxey. But by ambushing the transport, he also crossed Rothstein, who had allocated that scotch to Waxey. Feeling the pressure that came along with his large bootlegging business growing, Rothstein decided to step away from bootlegging in 1921 and focus on his gambling business. Waxey would inherit the bootlegging business.

From 1921 to 1925, Waxey's bootlegging business grew to be one of the biggest in the country. But it all came down in 1925, when the wife of one of Waxey's ship captains, Hans Fuhrman, tipped off authorities. She had complained to her husband's bosses that her husband came home from his trips drunk and out of money and begged them to send her part of his pay. They refused, so she ratted them out. In September 1925, Prohibition agents boarded the *Natisco* off Long Island and found four thousand cases of liquor, maps, radio codes and a list of customers. Waxey, who had been on a trip with his family in France and Germany, turned himself in on October 20. A week later, he, Max Greenberg and thirteen members of his gang were indicted and charged.

Despite being under around-the-clock police protection while waiting for the trial, Hans was shot and killed, with the official report claiming that it was suicide. Hans's wife and the New York Prosecutor's Office refute that report, but regardless, Waxey and Big Maxie walked.

The next few years saw Waxey focus on his other business, such as his gambling operations and real estate investing. Big Maxie, on the other hand, headed west, back to St. Louis in March 1921. There, he joined the Hogan Gang, as his old gang, Egan's Rats, refused to welcome him back. The next two years saw St. Louis become the scene of a gang war, which saw Greenberg, the main target, wounded in a drive-by and Willie Egan, leader of Egan's Rats, killed by the Hogan Gang in the fall of 1921. Greenberg inevitably left St. Louis and headed back east, where he met up with his old partner, Waxey Gordon.

Over the next few years, Waxey Gordon and Max Greenberg formed a formidable partnership. In 1928, Max Hassel entered the Jersey scene when the shutdown of his Reading, Pennsylvania breweries pushed him into buying the Camden County Cereal Beverage Company. That same year saw Dutch Schultz partner with Johnny Noe and open the Hub Social Club in the Bronx. The two quickly built out their business, eventually controlling most of the Bronx beer supply and moving into Manhattan. With Owney "the Killer" Madden, the owner of the Phoenix Cereal Beverage Company, the largest brewery in New York City, hesitant to do business with Dutch, the two turned to North Jersey to get their beer.[91]

In July 1929, Max Hassel signed a lease at the Carteret Hotel in Elizabeth, just outside Newark and across the bay from Jersey City. With Hassel now settled in New Jersey, the "Jersey Trio"—Hassel, Greenberg and Waxey—took over the North Jersey beer market. Together, the Trio owned somewhere in the neighborhood of seventeen breweries. But it didn't come easy, cheap or without huge risks.

Within months of Hassel's arrival, Frankie Dunn—a major player early in Prohibition—was killed. James "Bugs" Donovan, another early boss, had also announced his retirement. But as Max Hassel biographer Ed Taggert put it, "The reality of trying to quit the mob was brought home in dramatic fashion to Dunn on March 7, 1930." While Dunn had exited the bootlegging business, he hadn't quit the crime world altogether and instead shifted into a Rothstein-like role, financing various operations. Some of his old partners, though, weren't happy with how the finances had worked out when he left bootlegging. Dunn was gunned down as he entered his Hoboken office building. The Irish mob was growing weak.[92]

Meanwhile, the Jersey Trio was growing strong. Waxey had already established a large network of breweries in Union, Essex, Hudson and Passaic Counties by the time Hassel arrived. Running hoses through sewer pipes had become a trademark of Waxey's. Full-strength beer was transported through underground pipes to nearby buildings, so inspectors only saw the near beer that was legal at the time. It's likely, though, that agents were really bribed, not tricked. At the time, the salary of inspectors ranged from about $1,000 to $2,500 per year. That made bribery especially easy. Raids were also able to be avoided, though. On March 21, 1930, agents from Albany, New York, raided the Hensler Brewery, resulting in thirty-six arrests and seven thousand barrels of beer being seized. While the owners were never identified, the brewery had a pipeline running underground to a neighboring building, which seemed familiar.

Another raid took place on September 16, 1930, at the Rising Sun Brewery in Elizabeth, not far from Hassel's penthouse at the Carteret. While Mickey Duffy appears to have been the main operator of the brewery, it seems that the Jersey Trio was also involved. Duffy, of course, had been a partner with Hassel at the Camden County Cereal Beverage Company. While their relationship started out on the wrong foot, the two grew into a close partnership. When the Camden Brewery was shut down, Duffy sought revenge against John Finiello, who was already on his hit list. Finiello had once taken a $10,000 bribe to not raid one of Duffy's South Jersey distilleries but instead handed the money over to his boss and raided anyway.

Despite warnings to stay away from Rising Sun, agent John Smith, who had been poking his nose around the brewery, and Finiello, along with four other agents, raided the brewery. They were involved in an intentional hit-and-run as soon as they left Philadelphia, but it didn't stop them. While in the brewery's boiler room, the agents were overrun by a group of gunmen. While the agents were being disarmed, Finiello, who had been in another part of the brewery, entered the room. Nick Delmore, who worked at the brewery, yelled, "There's Finiello, get him!" Finiello was killed, and while several men were named as suspects in his murder, no one was ever prosecuted. Duffy, though, did see his rank rise on the feds' target list, which was partially the reason the Jersey Trio never let him become a full partner. He was killed less than a year later, on August 29, 1931, when he was shot three times in the head. One leading theory was that it was Duffy's own men who did the hit, although another is that Paul "Dago" Corbo, hired by the Trio, did it.

With Duffy gone and several members of his gang—including Samuel Grossman and Albert Skale—following, the Jersey Trio continued to grow stronger. But with the heat that came from the killings, Hassel was growing cautious and rarely left his room at the Carteret. In fact, in July 1931, he signed a new lease and expanded his suite to include the entire eighth floor, although there was only one way to enter it: room 824.

Around this same time, Dutch Schultz was growing more powerful in New York. Since he was buying so much beer from the Jersey Trio, he demanded to be included as a partner but was refused. Schultz responded as Schultz always did, with force. For a time, there was quite a bit of violence, as Schulz's crews raided the Trio's shipments. Things calmed down eventually, until Dutch learned that the beer he was getting from Waxey and company was a lower grade than what Hassel's favored Philadelphia distributors were getting. Dutch demanded the better product, but despite the fact that Hassel was known for diplomatically avoiding violence, he refused.

Throughout this time, the early 1930s, tensions with the Big Six and Dutch were growing. The "War of the Jews" was still ongoing between Waxey and Meyer. Hassel was cautious and rarely left his suite. Things were made worse when, on March 23, 1933, the body of Hassel's close friend, Al Lillien, was found with three shots to the head. Finally, Hassel turned away from his nonviolent ways and began taking a bodyguard with him.

At this time, Prohibition was on its way out. Unlike most gangsters, Hassel was eager for Prohibition to end. He looked forward to making an honest living, when he could become a legitimate brewer and not have to worry about federal raids and threats from competitors. In March, Hassel began doing favors for William Egan, a member of the state's Beer Control Commission, so that a month later, when 3.2 percent beer became legal, he'd be able to brew. While people with criminal records were unable to get licenses, Hassel's record was clear because he had settled all his cases. His partners weren't able to get licenses, but Hassel was able to get a single permit for all of their breweries.

The Trio's breweries received a huge influx of orders leading up to 3.2 percent beer. Once it hit the market, starting with the Harrison Brewery's 3.2 Old Heidelberg, they could hardly keep up with orders. Distributors that had gone through Schultz were now going directly to Hassel. Schultz wanted a discount on the product that he purchased but was turned down. It was clear that Hassel, as he had wanted to do for many years, was going to run a legitimate business. The Big Six didn't like that, and they didn't like that their offer for a partnership was turned down. Waxey didn't like it either; he knew that he'd have to be a background player in the legal operation due to his criminal record.

At 4:30 p.m. on April 12, 1933, Max Hassel and Max Greenberg were shot and killed in Hassel's suite at the Carteret. There are still many questions about the events that took place on that day. Among them is the movements of Waxey, who had been with them to discuss business. Waxey's story was that he had moved to room 804, where he enjoyed the company of Nancy Presser, the mob's prostitute. He denied that he moved into the other room to avoid the gunfire that was soon to ensue. Another odd movement belonged to Lou Parkowitz, Hassel's bodyguard, who claimed that he left room 824 for a few minutes a little after four o'clock. That Hassel's bodyguard was missing during the shooting was naturally suspicious.

Theories of who did the hit and why were many. Frankie Carbo was one suspect, as he was reportedly seen at the hotel that day. Carbo was a hit man for Murder Inc., possibly hired by the New York mob to eliminate the men who stood in the way of their post-Prohibition plans. Or perhaps he was hired by Waxey, who was not in a good position to profit from Hassel's legal operations. Carbo was arrested for the murder years later but was released for lack of evidence. Another theory has it that Greenberg was the real target—he was known for ripping off customers and then bragging about it—and that Hassel was killed simply because he was in the room at the time.

Within hours of the murder, the Jersey Trio's breweries in Paterson, Harrison, Union City and Newark were wiped clean. Again, there's no lack of theories

New York City Police Department photograph of Waxie Gordon (left) and two of his associates. *Library of Congress.*

about who it was. It may have been Waxey, cleaning up before the police came around. It may have been Schultz, who knew that the Trio's gang would be in hiding. Or it may have been the New York mob, which, having killed two of the three men, "sent their trucks to pick up the spoils of war."

With Hassel and Greenberg gone, U.S. Assistant Attorney Thomas Dewey—the man Dutch Schultz wanted dead—focused his tax case on Waxey. As part of the "War of the Jews," Meyer Lansky had his brother, Jake, leak information to the government about Waxey's beer operations. After hiding out in a Central Park West apartment and then in the Catskill Mountains, Waxey was arrested. Dutch Schultz also ran from a tax case. Still, even with Waxey in jail and Schultz on the run, the two gangs continued their gang war, with a total of thirty-four dying.[93]

Waxey was indicted on April 27, 1933. Throughout the trial, Waxey claimed to have been a minor player working for Max Hassel and Max Greenberg, whose operations were more widely known since police seized documents from Hassel's safety deposit box. Waxey claimed to have only been paid $300. The jury believed Dewey, who was able to prove that it was indeed Waxey's operation. He was fined $80,000 and sentenced to ten years. He was released seven years later, in 1940, on good behavior. But after promising that "Waxey Gordon is dead—that's all over. It's Wexler I'm interested in," Waxey found himself in trouble again, first for selling black market sugar that violated World War II rationing laws and again for selling heroin. In 1951, as he was arrested, he pleaded with Detective Sergeant John Cottone, the arresting officer, "Please kill me John—shoot me. I'm an old man and I'm through. Don't take me in for junk. How else can I live? Let me run, John, and then you shoot me." He was sentenced to twenty-five years to life at Alcatraz, where he died on June 24, 1952.[94]

AND SO, THE LAST of the Jersey Trio was dead, followed several years later by the other Jersey beer baron, Longy Zwillman. The Great Experiment was over, and beer was legal again in the Garden State. Many years later, in the fall of 2001, an interesting story would emerge in the September issue of *Gentlemen's Quarterly*. Joe Stassi, an old gangster, now ninety-five years old, took an interview with Richard Stratton in which he all but admitted to killing his best friend, Max Hassel.

Stassi wasn't a well-known name among mob historians, but he was certainly a player back in the day. He wasn't closely affiliated with any single Mafia family, but he did some work for some. In his first interview, Stassi recounted his stories of the origins of organized crime through to the 1960s. He told of his closeness with Hassel and with many other big names, but didn't tell too much of his own involvement.

Months later, he called Stratton so that he could tell him the rest of the story. He started by admitting his involvement in the assassination of Schultz: he was hired by Longy Zwillman and Meyer Lansky to kill Dutch before Dutch killed Dewey. He passed the contract along to Charlie Workman and Emmanuel Weiss, who carried out the assassination.

As the interview progressed, Stassi began to tear up as he recounted a job he received years before the Dutch hit: he was ordered to kill his best friend. He would hint that the friend was Hassel, but it was Hassel's murder that Joe wanted to discuss. He noted that in April 1933, he was called in by the New York mob, where he was ordered to kill Max Hassel and Max Greenberg.

He remembered questioning the order, telling his bosses that Hassel wasn't a threat. He remembered talking to Hassel earlier. Max asked who he thought would win if there was a war. Stassi told him that no one wanted war with Hassel. Greenberg, maybe, but not Hassel.

But the New York mob wanted both gone, and Joe was the man to do it. He was among the very few who could get into room 824 with a weapon. Although it was clear that Stassi admitted to killing Max Hassel, he refused to identify him as the best friend in his story.

In December 2001, Ed Taggert met with Joe Stassi and again asked him about the Max Hassel murder. He answered detailed questions about the hotel room where the murders took place. He said that Waxey was in the room when it happened, giving evidence that Stassi was there. He told about how people in the room reacted when shots were fired. "You must have been in the room to see them." Taggert said. "That's what the cops told me," Joe responded with a smile.[95]

CHAPTER 7

AND THEN THERE WAS ONE

1933–1994

T he wars between the various mob factions proved to be beneficial to legitimate brewers. Had the mob remained strong in North Jersey, it stood to hold on to its power even once Prohibition came to an end. But the wars and tax evasions caught up to it, allowing legitimate brewers to enter back into the market in 1933. The next few years saw old breweries start up again, as well as some new ones form. Some breweries, like Krueger's and Ballantine, saw huge success over the next few decades. Unfortunately, the decades following Prohibition ultimately became defined by consolidation, which in the end saw only one brewery operating in the Garden State: Anheuser-Busch.

One of the early struggles small breweries had to deal with was the rapid change in technology that had occurred while they were out of business. After thirteen years of not being able to go to the local pub, the population wasn't ready to renew its demand for draft beer. Around this time, home refrigerators were becoming more common. The result was that people wanted to buy beer and bring it home, meaning brewers had to bottle their beer. This wasn't something every brewery could easily do, and only the bigger brands were able to adjust.[96]

In New Jersey, there were a few notable breweries that were able to open back up in 1933, ready to compete. The first was Krueger's, which—despite the death of Gottfried Krueger in 1926—had weathered Prohibition by producing near beer and was therefore the only Jersey brewery ready to serve immediately after Prohibition had been lifted. Beginning at midnight on April 7, 1934, breweries were allowed to legally sell beer that was no more than 3.2 percent alcohol, and that's exactly what Krueger's did—it sold its beer right from the brewery's doorstep. Newspapers reported that it took two days for crowds outside the brewery to settle down. It's also reported that about 31 million gallons of beer were sold on that first day.[97]

Krueger's also had another first: it was the first brewery to ever can its beer. While cans were commonly used at the time to store food products, they hadn't been used for beer for two reasons. First, cans at the time couldn't withstand the high pressure that carbonated beer produced. Second, beer had a bad reaction with the tin that cans were made of. Following Prohibition, the American Can Company solved these two problems by strengthening the seam and coating the inside of the can.

Still, American Can needed a brewer to test the new product, so it made Krueger's, a well-established, reputable brewer, an offer it couldn't refuse: American Can would install the canning equipment for free, and Krueger's would only pay for it if it was successful. Still struggling from the effects of Prohibition, Krueger's agreed to try it. In 1933, the equipment was installed, and a two-thousand-can test run was done. Krueger's, nervous that the cans might explode or the beer might taste different, released the test cans only to regular Krueger's drinkers, of which 91 percent said they liked the new product. On January 24, 1935, Krueger cans went on sale to the public for the first time in Richmond, Virginia. When that test run proved to be a success, Krueger's committed and began selling the cans everywhere.

The gamble paid off for American Can. Later that year, Pabst joined the canning movement, originally only canning its export beer before including its popular Pabst Blue Ribbon. By the end of 1935, thirty-seven breweries were canning beer, although some had to include phrases such as "12 oz Same as bottle" to assure customers that they weren't missing out on any beer.[98]

Over the years, the canning industry saw a few experiments and many developments. Continental Can, for example, made a can shaped like a bottle. Not only did this reinforce the fact that the cans held the same volume of beer, but it also allowed the cans to be filled on bottling lines that many breweries already had. In the 1970s, pull-tab cans replaced flat-top cans and eliminated the need for a church key, which was used to punch two holes in the top of the can.

While the canning trend caught on throughout the late 1930s, it was halted with the outbreak of World War II, when tin was redirected to the war efforts. This helped the producers of bottles, which had (unsuccessfully)

Krueger Beer and Ale. *Bart Solenthaler, Bart & Co.*

attempted to convince consumers that cans gave beer an off taste. While bottle caps (or crowns, as they were widely called) were in short supply, a recycling effort helped to prop up supply.[99]

In addition to being the first to can its beer, Krueger's—which dropped the *s* in the late 1930s to become Krueger—also became known for its large advertising campaigns, the most famous of which was the "K-Man" logo. Not only was K-Man included on Krueger's first cans, but a giant neon sign of K-Man, fifty-seven feet tall and fifty feet wide, was also built in 1937. Over the next two decades, Krueger grew to become one of the biggest breweries in the country, reaching the 1-million-barrel mark in 1952.

Unfortunately, the 1950s saw the beginning of consolidation in the industry, as the macrobrewers like Anheuser-Busch and Miller began hogging market share. In 1961, the Krueger brand was sold to Narragansett Brewing of

Cranston, Rhode Island. Four years later, in 1965, Falstaff Brewing acquired Narragansett, and the Newark plant was shut down. An antitrust lawsuit eventually forced the brand into the ownership of Pabst, leaving Krueger beer as just a memory.[100]

The other major Jersey brewery to come back after Prohibition was Ballantine, though not without its changes. Prohibition had caused the brewery to lose much of the expertise that it had developed, so the old management decided to part ways with the brewing industry. Carl and Otto Badenhausen bought the brewery and, with the help of Scottish brewmaster Archibald MacKenchnie, rebuilt the Ballantine brand.

That brand continued to build over the next few decades. Following Krueger, Ballantine was one of the first breweries to join the canning movement and was one of the first to sell cans in six-packs to be purchased and brought home. Throughout the 1940s and 1950s, Ballantine was a sponsor of the New York Yankees, leading to announcer Mel Allen calling a Yankee home run a "Ballantine Blast."

By 1950, Ballantine was brewing 4.3 million barrels per year, making it the third-largest brewer in America behind Schlitz and Anheuser-Busch. Among fans of the beer were Marilyn Monroe, Frank Sinatra, Joe DiMaggio, John Steinbeck and Ernest Hemingway. While Ballantine XXX Ale was the most popular of its beers, the brewery also produced lager, porter, stout and an IPA. Its Burton Ale was also popular beer, although it was never sold commercially. Instead, the beer—which was brewed in small batches and aged in oak for ten to twenty years—was given as a holiday gift. Bottles were labeled as "Special Brew: Not For Sale" and were printed with "Brewed Especially For" followed by the recipients name and the dates it was brewed and bottled.[101]

The Ballantine brand gained a big boost in 1949. When brewery workers decided to strike, all the breweries in New York City were forced to close, as were the taverns and pubs that found themselves without beer. Fearing what might happen next, the labor union requested that Ballantine ship beer to the city, which it hesitantly agreed to do. On April 15, 1949, Ballantine, along with Krueger, began delivering ten thousand barrels of beer to the city. Worried that customers might get used to these Jersey beers, the city's breweries and workers came to an agreement. Along with Krueger, Ballantine helped end the strike.[102]

By 1960, despite attempts to reinvent its recipes and revitalize the brand with new advertising, Ballantine sales fell considerably, and Ballantine fell to sixth in the brewery ranks. By 1965, the company was losing money, and

Ballantine—America's largest-selling ale. *Bart Solenthaler, Bart & Co.*

in 1969, the Badenhausen's sold to Investors Funding Corporation of New York for $16.3 million. But without any brewing experience, the group of bankers was unable to fix the problems that now caused Ballantine to fall out of the top ten. In 1972, just three years after purchasing the company, Investors Funding sold the Ballantine brands and distribution network to Falstaff Brewing for $4 million plus royalties. It retained the physical brewery site, with hopes of developing it for other uses, but that, too, failed. It declared bankruptcy in 1974.

Falstaff produced Ballantine beers at its Cranston, Rhode Island brewery for three years, but after losing about $22 million, it sold it to Paul Kalmanovitz, who also owned Pabst. Kalmanovitz cut production and advertising, which, despite improving the brand's profitability, made Ballantine hard to find.

The production of Ballantine later moved to Fort Wayne and then to Pabst facilities in Milwaukee. When that site was closed in 1996, Ballantine, like Krueger, became just a footnote in the history books.[103]

Flash-forward to 2014: Ballantine IPA is back, thanks to Pabst brewmaster Greg Deuhs, who years ago began thinking about ways to resurrect that famed beer. With all the changes in ownership, no useful records of how Ballantine was produced survived. Using what little he had, he combined his and others' memories and some educated guesses to resurrect that famed IPA.[104]

Two other pre-Prohibition breweries also made an attempt to join the new order. The first was the Peter Doelger Brewery, which came from the marriage of Karl Bissel—grandson of Peter Hauck—and Phoebe Doelger in 1917. While Prohibition saw the brewery fall into the hands of Max Hassel and Max Greenberg, who renamed it the Harrison Beverage Company, the Doelgers regained control after Prohibition. Despite rereleasing some of its popular brands and creating a "Half & Half" lager-ale hybrid beer, Doelger couldn't keep up with the consolidation that came about following World War II. It closed in 1947.[105]

Finally, there was the Orange Brewery, which had been around since 1901 but stopped brewing during Prohibition. In 1934, it was sold to John Fr. Trommer Inc., which sold it again in 1950 to Liebmann Breweries of Brooklyn. Orange Brewery was used as the bottling facility for Rheingold, which Liebmann owned, until it was shut down in 1977.[106]

The post-Prohibition era also saw the emergence of new faces in the New Jersey brewery landscape. The first was Pabst Brewing Company, an already established brewery from Milwaukee known for its famous Pabst Blue Ribbon brand. In 1945, Pabst bought the Hoffman Brewery in Newark. Hoffman brewed beer but was known more widely for its soda. Hovering above the Hoffman brewery was a sixty-foot-tall, fifty-five-thousand-gallon water tower shaped like a bottle. When Pabst purchased Hoffman, the bottle was painted blue but slowly turned red from rust. The Pabst facility was closed in 1982. Despite attempts by a local preservation group to have the seventy-five-year-old bottle declared a landmark, it was taken down in 2006 when the building was demolished.[107]

And finally, the big one: Anheuser-Busch (A-B). In the mid-1940s, A-B, already a large brewer in St. Louis, Missouri, purchased fifty acres in Newark on which it built a large, modern brewery. The Newark facility opened in 1951 to help get fresh beer to the large New York City–area market. Fresh, quality beer is important to A-B, and it always

Pabst Blue Ribbon was once a Newark creation. *Bart Solenthaler, Bart & Co.*

has been. It dates back to the brewery's founder, Eberhard Anheuser. After Anheuser married the daughter of a Bavarian brewer, he joined the business in 1865. He went on to travel throughout Europe and America, tasting beers and studying the art and science of brewing. To him, using the highest-quality ingredients was required if he wanted to brew the highest-quality beer.

In 1954, another brewery was opened in Los Angeles. By 1957, A-B was the biggest brewery in the country, a status it still holds today. Today, Anheuser-Busch operates thirteen regional breweries for the same reason it expanded to Newark: to ensure that every market gets the freshest beer possible.[108]

North Jersey Beer

In Newark, brewmaster Jorge Garcia is responsible for brewing more than 7 million barrels per year of Budweiser, Bud Light, Michelob ULTRA, Busch, Busch Light, Natural Light, Natural Ice, Budweiser Select and King Cobra for the New Jersey, Delaware, New York, Pennsylvania, Rhode Island and Connecticut markets.[109]

PART III

THE CRAFT BEER REVOLUTION

CHAPTER 8

RECOVERY

1994–2000

At the turn of the twentieth century, New Jersey was one of the biggest producers of beer, and Newark was the beer capital of the country. Names like Ballantine, Krueger and Feigenspan were recognized around the United States. But Prohibition and industry-wide consolidation destroyed New Jersey breweries. By 1984, there was only one brewery operating in New Jersey: Anheuser-Busch. That all changed when, in June 1994, Dave Hoffmann and his father, Kurt, opened Climax Brewing in Roselle Park. New Jersey craft beer was back. Two years later, Gene Muller turned Flying Fish Brewing Company from a virtual brewery to a real craft brewery that became, and remains, the biggest craft brewery in the state. In the meantime, a 1994 law legalized brewpubs, making way for the Ship Inn in Milford to become the state's first brewpub since Prohibition; it has since been followed by many more.

By 2000, New Jersey had reestablished itself as a serious beer state, especially when it came to craft beer. And it was North Jersey that was leading the way. Names like Climax, High Point, River Horse and Cricket Hill were recognized even past New Jersey borders. Meanwhile, brewpubs like the Ship Inn, Triumph, Long Valley, Harvest Moon, J.J. Bitting and Krogh's were becoming favorites among locals and visitors alike.[110]

Brewing tanks greet you as you enter Harvest Moon. *Chris Morris.*

Climax Brewing | Roselle Park

Dave Hoffman is famous in the New Jersey beer scene for many reasons. The most obvious is that he's the founder of Climax Brewing, the first modern microbrewery in the state of New Jersey. What most people don't realize, though, is that he's also one of the coolest brewers around. His passion for his beer in unmatched, and it shows.

Dave started off in the beer world as the owner of the Brewmeister, his homebrew shop in Cranford, New Jersey, and as a consultant for Gold Coast Brewing Company. When his father, Kurt Hoffman, had extra space in his store, Dave suggested installing a small brewery, and that's exactly what the father-and-son team did.

His father, after all, was the reason he drank quality beer. Born in Germany, Kurt was used to drinking the excellent beers that were commonplace in Europe. As Dave grew up, he began drinking his father's beers and found himself unable to drink the macro-produced beers brewed that made up the majority of the market in the United States. He began homebrewing to fill his own demand for quality beer, a hobby that would later lead to him starting the first microbrewery in New Jersey.[111]

The Climax brew house. *Climax Brewing*

The brewery—which was founded in June 1994 but didn't begin selling until February 1996—started off small; at first, all it had was a four-barrel brew house, four fermenters and a carbonation tank. The name, as Dave explained in an episode of Blip Network's *The Brewery Show*, has an interesting story as well. When Dave and Kurt were ready to open the brewery, they were searching for a name when someone suggested Climax. "I thought, 'That's an interesting word,'" explained Hoffman. "So I looked it up in the dictionary, and it said, 'the point of greatest excitement.' And I thought, 'What a stellar word. That's what we're gonna call our brewery,' because after a year of building this brewery, the point of greatest excitement for us was to open the door on that first day and brew our first batch of beer."[112]

That first beer—despite being raised on German beer—was an English-style extra special bitter (ESB), first brewed on February 27, 1996. It has since gone on to win many awards and accolades, not the least being recognized by famed beer writer Michael Jackson as one of the best beers in the world (Michael Jackson's books, including *The World Guide to Beer* and *Ultimate Beer*, are still regarded as the most influential beer books in the world). The ESB was followed by Climax's porter and then an IPA. The brewery has since

Several of Climax's offerings. *Climax Brewing.*

gone on to brew many others. The Climax brand of beers—which is all of the brewery's ales except for the hefeweizen—currently consists of the ESB, IPA, nut brown ale and cream ale, all of which are year-round offerings. The seasonal offerings fall within the Hoffman brand, and with the exception of the hefeweizen, are lagers; they include Hoffman Helles, Oktoberfest, doppelbock and Hefeweizen Wheat Ale.[113]

The brewery has grown since those modest four-barrel days. With a fifteen-barrel system designed by Hoffman, Climax currently produces about one thousand barrels per year, it but has the capacity to brew up to four thousand barrels. Unlike most modern breweries, which utilize expensive automated bottling lines, Climax uses a manual six-head counter-pressure bottling system also designed and built by Hoffman. It should come as no surprise that Hoffman also self-distributes his beer. As he told Lew Bryson and Mark Haynie, authors of *New Jersey Breweries*, "Nobody knows more about my products than I do. Besides, I'm already on the payroll." Perhaps the only thing Hoffman doesn't do its design the label artworks; that's done by Hoffman's longtime friend Gregg Hinlicky.[114]

Hoffman also has an appreciation for the history of beer in the Garden State, one that now includes him and his brewery. Speaking on the quality of New Jersey's water, Hoffman explained that the water in North Jersey was one of the reasons Newark was once home to some of the biggest and most successful breweries in the world. "We have some of the finest brewing water on the planet," Hoffman has said. He told Michael Pellegrino, author of *Jersey Brew: The Story of Beer in New Jersey*, that if "water is too hard then it is no good for lagers. If it's too soft then it is no good for ales. Our water is perfect for all styles." And it shows. Hoffman has been brewing award-winning, world-class beer for nearly two decades.[115]

HIGH POINT BREWING COMPANY | BUTLER

Like many brewery founders, Greg Zaccardi started off as a homebrewer. While earning a degree in chemistry at the University of California–Santa Cruz, Zaccardi was a member of the school's Good Beer Club. When he returned to the East Coast, he found the amount of quality beer a bit lacking. As he told Bryson and Haynie in *New Jersey Breweries*, "When I returned, the most exotic thing I could find was Molson Golden." He solved that problem just like many of New Jersey's other early brewery founders: he started brewing. "I realized that the only recourse was to learn to brew my own."[116]

His path to founding his brewery was helped out by more than just his homebrewing hobby, though. His soon-to-be wife came from a brewing family, with brewers still working at a brewery in Bavaria. After her father discovered his homebrewing hobby, Greg found himself in Bavaria, working as a brewing apprentice and learning about the traditional German beers styles and brewing methods. A year later, he brought this newfound knowledge back to the Garden State. "I loved the beer of Germany, and I wanted to be able to share them with Americans who may never get the chance to travel there," he told Bryson and Haynie.[117]

His first choice location for the brewery was an area called High Point in Wantage, New Jersey. The area—which is the highest point in New Jersey—reminded him of the Bavarian mountains, where he had developed his passion for German beers. The brewery, which was founded in 1994 and named High Point Wheat Beer Company—found its home in Butler, New Jersey, south of Wantage but close to a spring-fed reservoir that provided the soft water needed to produce the German-style lagers that High Point was soon to be producing.[118]

In addition to losing out on his first choice location, Zaccardi faced one more hurdle: he needed special brewing equipment that would help ease some of the problems that come while brewing wheat beers.

At High Point, Zaccardi began turning out award-winning beers under the Ramstein brand. The name Ramstein is named after a German city that was close to a U.S. Air Force Base and served as a home to a large American population. The name, he figured, was a good way to show the connection between his German beers made in America. He'd later describe these beers as "[t]raditional German beers made with American innovations."[119]

And German beers they are. For some time, High Point produced only wheat ales brewed with German grains and noble hops—Tettnanger, Hallertaue, Spalt and Saaz hops grown in certain European regions. Zaccardi even got the Bavarian brewery to allow him to use its proprietary yeast strain exclusively in the United States. High Point has since expanded to brew beers that do not include wheat, and the brewery dropped the wheat reference to become simply High Point Brewing Company.

High Point now operates a fifteen-barrel brew house and produces about four thousand barrels per year, which is right at the brewery's capacity. It also serves as the contract brewer for BOAKS Beer and Bolero Snort Brewery. (Contract brewing is when a larger facility, such as High Point, is hired by independent brewers, such as BOAKS and Bolero Snort, to brew their unique beers for them, since they don't own or operate their own facilities.) More on BOAKS and Bolero Snort later.

High Point has gone on to produce award-winning beer, recognized by nationally known competitions and beer critics. *All About Beer*'s Charles Finkel called Ramstein Blonde "one of the best American wheat beers I've tasted," while Charlie Papazian, president of the Brewers Association, called it "a willfully wonderful wheat worthy of wanting. I'm actually astounded." Michael Jackson, the famed beer writer, labeled Ramstein Classic Wheat one of the best dark wheat beers in the world. And Ramstein Maibock was once rated the best new beer on the popular beer rating website Beer Advocate.[120]

Greg Zaccardi founded High Point to introduce the East Coast to the wonderful German styles that were lacking back in 1994. As he put it, "My vision was always to make beergarden-fresh beer for the Garden State, and I believe I've succeeded in that goal." Indeed he has.[121]

THE SHIP INN | MILFORD

Not unlike New Jersey beer history in general, the Ship Inn of Milford boasts a rich history, one not lost on its owners or customers. The building that houses the restaurant and brewery dates back to the 1860s. It once served as a bakery and then as an ice cream parlor that included a speakeasy in the back that served alcohol during Prohibition. It was also a two-lane bowling alley in the 1950s and '60s and then a tavern. In 1985, David and Ann Hall, both of England, bought the tavern, then known as the Town Tavern, and began to feature their distinctive English cuisine.[122]

Then, in 1994, when New Jersey law was changed to allow for brewpubs, Ann Hall knew what was to come next: the Ship Inn would start brewing. Specifically, it would become the first modern brewpub in New Jersey. That's what happened on January 3, 1995, when the Shipp Inn brewed its first batch of traditional English-style ale, and that's what its been doing ever since.

The beer is perhaps the most unique thing about the Ship Inn, even with its great bar top, which is made from the wood of the bowling lanes that used to occupy the building. Keeping their English roots in mind, the Hall family, led by son and brewer Tim Hall, brews traditional English-style ales, cask and all.

Cask ale is the traditional way beer was consumed. Cask ales go through a process similar to beers that are bottle-conditioned (see chapter 2), where yeast is added to fermented beer (along with a little bit of sugar or wort) in the bottle, where it ferments, creating carbon dioxide and carbonating the beer. Instead, the beer is put into a cask along with some sugar or wort, where it is then sealed with a bung. Over the course of a few weeks (or longer), the yeast, which is still alive from the original fermentation, consumes the newly added sugars, producing a light carbonation before falling to the bottom of the cask. The bung is then taken out, and lightly carbonated beer is available to drink.

Over time, methods to quickly force carbonated beer into steel kegs that could withstand high pressure largely pushed cask beer out of the way—especially in the United States, where drinkers prefer beer that is more carbonated—but many breweries still do occasionally produce beers that are available in cask format, while few have extensive selections of cask beers. The Ship Inn, though, continues the traditional cask method. As brewer Tim Hall told Bryson and Haynie in *New Jersey Breweries*, "We do English beer…our beers are brewed to be cask ales."[123]

To get the brewery set up, the Halls approached Alan Pugsley, an expert in organizing English-style breweries. At the time, Pugsley was working with Englishman Peter Austin. Even today, the Ship Inn continues to use the seven-barrel Peter Austin brewing system. It also uses an English yeast strain that Austin had called Ringwood.

According to Hall, some people have a problem with Ringwood, which creates characteristics that some deem undesirable. "What's wrong with a yeast strain that imparts flavor?" Tim Hall asked Bryson and Haynie. "It's core to this type of beer. You work with Ringwood and it becomes your yeast, it fits to your house." Also unique to Hall's beers is the water, which comes from the Delaware River just a stone's throw from the brewery. "That's the water I've got, and I'm going to make the beer it lets me make."[124]

Not many are complaining. For nearly two decades, Tim Hall and the Ship Inn have been crafting English-style ales that would evoke pride even from his friends from London, where he grew up. Using the seven-barrel system, the Ship Inn turns out about four hundred barrels per year, with a capacity of just over five hundred. The bar's permanent fixtures include a golden wheat, an extra special bitter and a best bitter (British pale ale) and are generally available along with a porter, brown ale or stout, plus a seasonal.[125]

Triumph Brewing Company | Princeton

Many of New Jersey's breweries and brewpubs boast fascinating architecture and interior design, but none more so than Triumph Brewing Company in Princeton. When you walk in the front door, you enter a hallway—a very, very long hallway—that leads to the brewery and restaurant that sit off Princeton's popular Nassau Street. As you enter the bar and dining area, you begin to take note of the spectacularly designed brewpub.

You enter into a large dining area with thirty-foot-high ceilings. To the left is a staircase that leads upstairs to another dining area. To the right are stairs that lead down to the bar, behind which sit five large fermentation tanks enclosed in a glass case. Those tanks are involved in the brewing process that lets Triumph's Princeton location produce nearly 1,500 barrels of beer per year.

I specify the Princeton location because the original location (Princeton), which opened in March 1995, shortly after New Jersey began allowing brewpubs, was so popular that two more were opened in New Hope in

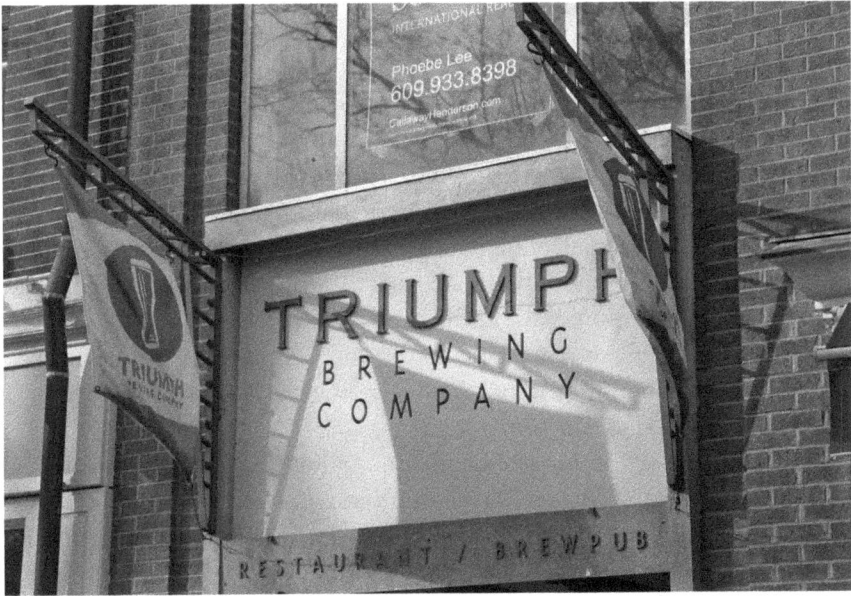

The exterior of Triumph on Nassau Street in Princeton. *Chris Morris.*

A look up at the Triumph fermentation tanks behind the bar. *Chris Morris.*

2003 and in Old City Philadelphia in April 2007 (the Old City location has since closed).

Triumph—which has the designation of being the first brewpub in New Jersey that started out all about the beer, not the food—is the brainchild of Adam Rechnitz, a former homebrewer. And it really was all about the beer, as Rechnitz stated in *New Jersey Breweries*. "I'd as soon have opened a brewery." His original plans had him serving as the brewer, but the responsibilities that came with him serving as owner forced him to delegate the job. Luckily for him, around the time he realized he couldn't serve as owner and brewer, he met Tom Stevenson, a homebrewing enthusiast who had been working as a horticulturist but was looking for a change. The rest, as they say, is history.

As Jay Mission, the former head brewer for the Triumph breweries, noted to Bryson and Haynie, Stevenson's non-brewing background was a good thing for Triumph, as it allowed him to add a different perspective; he was referred to by Mission as the brewery's "innovative brewer." It was this perspective that led to the creation of Triumph's Coffee & Cream Stout, Jewish Rye Beer and Gothic Ale, a modern version of beers that were brewed with a mix of herbs before the use of hops.

Today, Triumph's Princeton location boasts an impressive lineup of beers. A few regulars can usually be found on tap, such as the honey wheat, Bengal Gold

The tap list at Triumph Brewery in Princeton. *Chris Morris.*

IPA and amber ale, alongside award-winners Czech Pilsner (which won silver at the 2005 Great American Beer Festival) and Ruchbier (which won silver at the 2004 GABF). The Great American Beer Festival is the most prestigious brewing competition in the United States; the three Triumph locations have won a total of twelve medals at the Great American Beer Festival. Other favorites include the Coffee & Cream Stout, extra special bitter, imperial stout, Oktoberfest Marzen and Vienna lager. In total, the brewery has produced more than fifty different beers.[126]

Long Valley Pub and Brewery | Long Valley

The property that now houses the Long Valley Pub and Brewery dates back to 1771, when a German immigrant named Trimmer built a large barn that showcased stone walls and large wooden beams. The barn was closed with increased regulation on dairy farms in the 1940s, and the property was sold to Jack Borgenicht in 1965. In January 1995, work began to transform

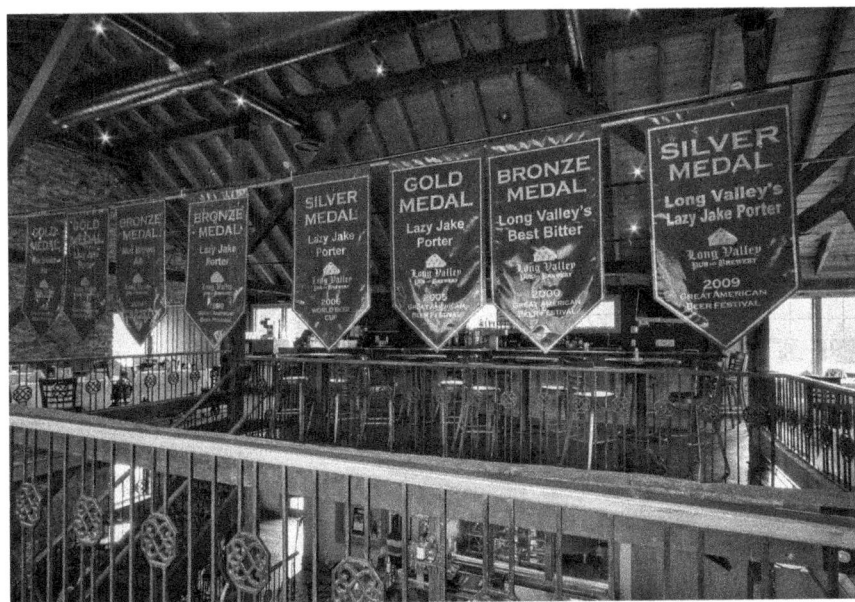

Banners honoring Long Valley Brew Pub's award-winning beers hang in the dining room. *Robert Greco.*

the barn into a brewpub, with the goal to keep as much of the original craftsmanship as possible.

The brewpub opened in October 1995 and immediately gained notoriety for its beers. In 1996, Joe Saia joined brewmaster Tim Yarrington in the brew house and has since taken over as brewmaster, a role he takes very seriously. "I have a responsibility to my customers to brew a consistent menu of products," he said for *New Jersey Breweries*. "Consistency is the key!"[127]

In 1999, Long Valley entered the Great American Beer Festival for the first time and won bronze with its Lazy Jake Porter. Lazy Jake won gold in 2000, while its best bitter won bronze. In total, Long Valley has medaled seven times at the country's most prestigious brewing competition, with its most recent coming in 2009.[128]

Saia is still responsible for all brewery-related operations, which includes maintaining the four Long Valley regulars: Long Valley Light, Golden, Amber and Porter. The brewery uses a seven-barrel system and brews right around the 750-barrel capacity.[129]

RIVER HORSE BREWING CO. | EWING

While River Horse is still a relatively young business, it's already gone through several major chapters of its history. The brewery was founded in Lamberville in April 1996 by three brothers—Jim, Jack and Tim Bryan—who decided to take their homebrewing hobby to the next level. Starting in the building of an old cracker factory that came with the right draining, River Horse had a promising start, but it quickly became just another brewery. As Lew Bryson wrote in *New Jersey Breweries*, River Horse beers were "solid, clean beers"—not exactly a compliment. "I never had a bad River Horse beer, but there weren't many that excited me either."[130]

That began to change when new owners took charge in 2007. Chris Walsh and Glenn Bernabeo had worked together as partners at SSG Capital Advisors, where they specialized in selling distressed companies. In 2006, they sold SSG to a Cleveland bank and found themselves looking for a new challenge. They decided that they wanted a change: "When we sold SSG in 2006, we were looking for another challenge. We wanted to make something, something that meant something to both of us." They wanted to run a company that made something they both cared about. They found that "something" when River Horse went up for sale.[131]

River Horse Special Ale American Amber. *Chris Morris*.

Despite being the second-largest brewery in the state, River Horse has plenty of room to grow at its new Ewing location. *Chris Morris*.

For Walsh and Bernabeo, it turned out to be the perfect product. The two were longtime fans of craft beer, although they weren't experienced brewers themselves. Still, they saw potential in the craft beer market. "We looked at the macro forces acting on the industry, and they're great," Bernabeo told Bryson. "Craft beer is a long-term opportunity. The consumers are wising up to quality; they're appreciating finer beer. They won't go back."[132]

They took over in 2007, and things at River Horse began to change, though not drastically. Rather than create a new brewery, they decided to build on what River Horse was already doing right, which was making good beer. Some things did change, though. Among the changes was a new location: they moved into a new brewery in Ewing, expanding from ten thousand to twenty-five thousand square feet, in April 2013. At the new location, fan favorites like Triple Horse (Belgian triple), Hop Hazard (American pale ale), Hop-a-Lot-Amus (double IPA), Belgian Freeze (seasonal dark Belgian) and Summer Blonde (American blonde ale) are brewed on a twenty-barrel system and fermented in one of eleven forty-barrel fermenters (two batches are brewed to fill each fermenter).

But changes are still coming. While the brewery currently produces about thirteen thousand barrels per year—making it the second-largest brewer in the state behind Flying Fish—the new facility has enough room to expand capacity to eighty thousand barrels. The brewery also started a hop garden outside the brewery in the spring of 2014 and plans to release a fresh hop harvest beer within the next few years (hops need several years to develop). The brewery recently released a new IPA that it expects to become its best seller and is continuing to develop its barrel-aged series.[133]

HARVEST MOON BREWERY & CAFÉ | NEW BRUNSWICK

New Brunswick has been home to a handful of breweries throughout its history, including the short-lived Thomas Teneson Brewery (1874–75), Rock Spring Brewery (1898–1910) and possibly a brewery owned by George Wiedenmayer before he moved back to Newark to start George W. Wiedenmayer Brewing Company in 1880. There was also New Brunswick Brewing Company, which has records of existing as early as 1907. The brewery returned to the scene in 1935 but wound up being one of many post-Prohibition breweries that had quick lives, closing in 1938.[134]

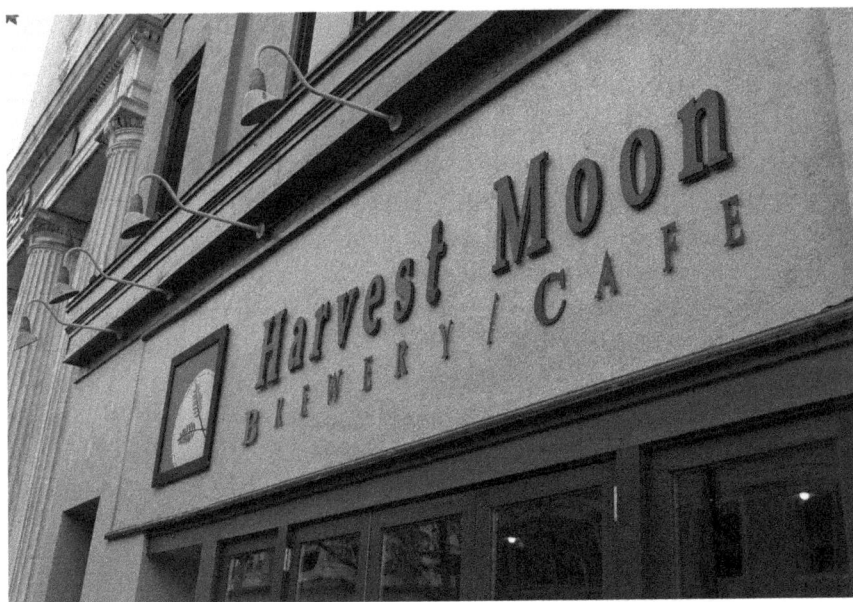

The entrance to Harvest Moon on George Street in New Brunswick. *Chris Morris.*

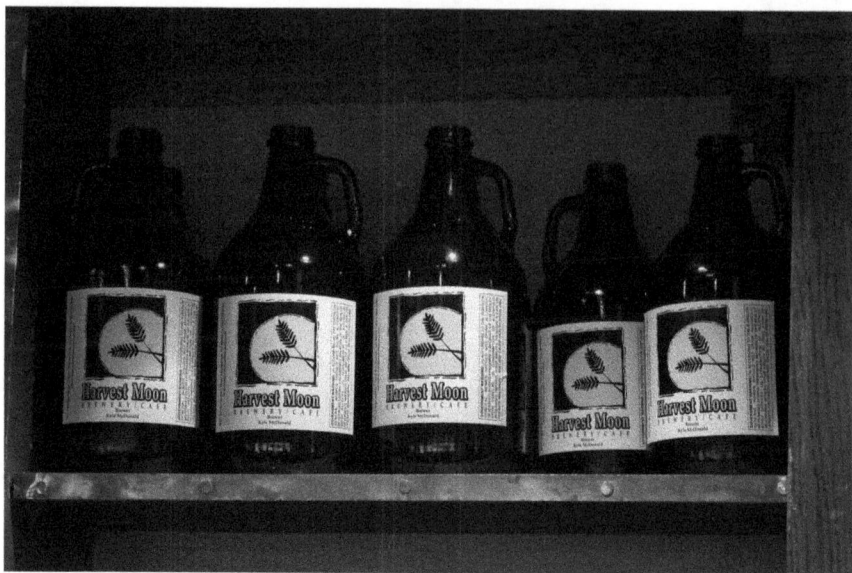

Harvest Moon growlers on the shelves behind the bar. *Chris Morris.*

For decades, the failed New Brunswick Brewing Company was the last brewery in the city of New Brunswick, which is surprising considering the city's location near major highways and train lines that lead to New York City, as well as its proximity to the Raritan River and the Raritan Bay.

That changed in 1996, when Harvest Moon was founded on New Brunswick's George Street in the heart of Rutgers' New Brunswick campus and a block away from the Johnson & Johnson world headquarters. For many years, head brewer Matt McCord crafted beers designed to appeal to the masses while remaining fresh and flavorful. He saw an interesting opportunity as the head brewer of the only brewery in such a large college town: "Being in a college town…filled with younger crowds that live off of commercial beer, it is my job to be a transition brewer…to transition the common beer drinker to the real flavor of beer, promote better beer and respect beer of any style." McCord is no longer the head brewer at Harvest Moon, but current brewer Kyle McDonald is carrying on the tradition well.[135]

McDonald got his first taste of craft beer during a six-week exchange program in Germany in 1996. The next year, just a senior in high school, he wrote a fifteen-page thesis on German brewing. After graduating, he began working in the kitchen of an Iowa brewpub; after months of pleading with the brewing staff, he made the switch into brewing with one shift of keg cleaning

Fermentation tanks in Harvest Moon's basement. *Chris Morris.*

per week, followed by a small promotion that saw his responsibilities expand to racking (transferring beer between vessels), carbonating and delivering kegs. Kyle eventually came east and was hired as an assistant brewer at Harvest Moon. When McCord decided to find a job closer to home, Kyle assumed the head brewer job in May 2010.[136]

On the brewery's ten-barrel system, Kyle continued to brew the beers for which Harvest Moon was known, though not without introducing some of his own unique ones. Among the more than fifty beers Harvest Moon has offered over the years are favorites such as Full Moon Pale Ale, an Irish Red, Elmes' Mild Manor, Moonlight Kölschbier and Galactic Amber, alongside a long list of creative Belgians and seasonal offerings like the Oktoberfest, Winter Warmer and imperial stout. One of the biggest favorites among New Brunswick beer drinkers was a beer developed by McCord in memory of Deputy Chief James D'Heron of the local fire department, who lost his life in the line of duty in September 2004. To honor the fallen hero, McCord brewed up Jimmy D's Firehouse Red Ale, with a portion of all sales earmarked for the Children's Burn Camp of the Connecticut Burns Care Foundation. A decade later, Jimmy D's Firehouse Red remains a brewery favorite.

The brewery also hosts a monthly cask night, where Kyle and company prepare a special cask beer for the local beer enthusiasts. Cask night beers

The bar of Harvest Moon. *Chris Morris.*

have included a Belgian Winter Ale brewed with dates and figs, a Maple Coffee Pumpkin Stout, a Blood Orange Lemon Sorachi Saison and a Cantaloupe Gose, among many others. The brewery also plays host to a local homebrew club.[137]

J.J. BITTING BREWING COMPANY | WOODBRIDGE

You can't talk about J.J. Bitting (commonly called "J.J.'s" or "Bittings") without talking about two things: the history of the building and the beer. Let's start with the building.

Like so many of New Jersey's breweries (and especially brewpubs), the building that houses Bittings has a history that dates back long before the brewery itself. The building at the corner of Main Street and Eleanor Place was built in about 1915 and was home to the J.J. Bitting Coal and Feed Depot and its own rail siding until the 1950s. When it was sold, it housed a furniture store, an appliance retailer and others until it closed in 1962, at which point it sat idle for more than three decades.

A train statue sits outside J.J. Bitting in Woodbridge. *Chris Morris.*

That's when Mike Cerami came in. After seeing brewpubs on the West Coast, Cerami, who had previously owned his own restaurant, knew that he wanted to open one in New Jersey. At first, he didn't even realize that they were legal here, but that didn't stop him. He began doing his research, and when he found out that he could open one, he filed the paperwork immediately. In February 1997, he opened J.J. Bitting Brewing Company.

The railroad that runs alongside the brewery is now a commuter train, part of the New Jersey Transit system. Still, the brewery pays tribute to its history. On your way in, you pass a statue of a train. When you enter, the area that used to be the loading dock is now a large, open dining room. In the middle, set behind glass, sit four ten-barrel copper fermenters, which are fed wort (unfermented beer) from the brew house that sits on the third level.[138]

Unlike other brewpubs that constantly experiment with new styles and recipes, Bittings offers a consistent offering of beers brewed by head brewer Mehmet Kadiev. Among the usuals are Avenel Amber, Blackjack Stout, Victoria's Golden Ale, Barely Legal Barleywine, J.J.'s Nut Brown Ale and W.H.A.L.E.S. IPA, named after the Woodbridge Homebrewers Ale and Lager Enthusiast Society, which J.J. Bitting hosts. While there may not be a new offering every week, one thing is for sure: the beer it does make is consistently good. Very good. So good, in fact, that it was once named the second-best brewpub in America by the Beer Advocate website.[139]

With its history so deeply rooted in Woodbridge and New Jersey, its no surprise that Bittings is also extremely involved in the community. Most notably, the brewery has sponsored the Central Jersey BeerFest for the past seven years. With beers coming from breweries around the state—including Tun Tavern, Pizzeria Uno, Harvest Moon, Crick Hill, BOAKS and the Ship Inn—as well as beyond New Jersey, the annual event raises about $10,000 to help families with children facing illness.[140]

TRAP ROCK BREWERY & RESTAURANT | BERKELEY HEIGHTS

Berkeley Height's Trap Rock Brewery & Restaurant opened in April 1997 as the eighth brewpub in New Jersey—after the Ship Inn, Triumph, Long Valley, Harvest Moon and J.J. Bitting, as well as Basil T's and Artisan's in Red Bank and Toms River, respectively. Trap Rock was the first of what is now ten restaurants operated by its parent company, Harvest Restaurants. The building, designed by Morris Nathanson, is designed to replicate a cozy European ski lodge atmosphere, boasting high ceilings, large windows, a large oak bar and a stone fireplace. More importantly, it houses a seven-barrel brewing system, which Charlie Schroeder has used to craft some of the finest beers in the state for nearly thirteen years.[141]

Upon opening, Trap Rock quickly gained recognition for its high-end food—something not commonly found at brewpubs—when it earned an "excellent" rating from the *New York Times*. But to appease the local customer base, the food was toned down a bit in the years following. Still, the food's reputation is topped only by its beer. That's largely because of head brewer Charlie Schroeder, a Siebel Institute– and American Brewers Guild–trained brewer who did his apprenticeship at Flying Fish and worked for six months at Pennsylvania's Victory Brewing Company before taking over for Rob Mullin at Trap Rock.

Just as the restaurant is consistently putting out new, innovative dishes to keep patrons happy, so, too, does the brewery put out an ever-changing menu of beers. Ghost Pony Helles Lager has long appealed to the masses, and it earned a bronze medal at the prestigious Great American Beer Festival in the 2001 judging of Munchner-style helles beers. Other beers—such as Abbey Du Roc, Cherry Stout and Polar Vortex Porter—push the envelope a bit more alongside beers like El Matador, a saison brewed with Serrano pepper and mango.[142]

Pizzeria Uno | Metuchen

There are a few things that New Jersey is famous for. Among Bruce Springsteen, the Jersey Shore and the age-old questions "Does Central Jersey exist?" and "Is it pork roll or Taylor Ham?" Jersey is known for its pizza—a slice of Jersey thin crust is simply unbeatable. Drive around any town in the state, and you'll see pizzerias galore.

Why, then, are there no pizzeria brewpubs, like the dozens that are scattered across the West Coast? This is a question a lot of Jersey residents might ask, but only because they don't know that the Pizzeria Uno, which so many drive by on the commute up and down Route 1, is actually a brewpub. According to its zip code, the brewpub falls in Metuchen, although geographically it's actually in Woodbridge. Pizzeria Uno, the famed deep dish–style pizza chain restaurant, originally planned to have a number of its restaurants brewing beer in-house and supplying other restaurants nearby, but restrictive laws prevented those plans from ever happening. They were left with just one brewpub, and its all ours.

In the middle of Menlo and Woodbridge Malls, two of the Garden State's biggest shopping centers, Pizzeria Uno Chicago Grill & Brewery was founded in March 1998, where head brewer Mike Sella quickly made a name for the brewpub with beers like 32 Inning Ale, Bootlegger Blonde and Station House Red Ale. Mike has since moved on and was replaced by Chris Percello, who was then replaced by Zac Conner in 2013. On a fifteen-barrel Pub Brewing system, Zac currently brews up beers like Gust'n Gale Porter, Ike's IPA, Coffee Stout and a popular Oktoberfest.

The brewpub also serves as host to a seasonal event it calls Cask Fest. Along with other Jersey breweries like Kane (Ocean), Carton (Atlantic Highlands), Bolero Snort (Ridgefield Park), Harvest Moon (New Brunswick) and Flying Fish (Somerdale), several specially brewed cask beers are tapped, and the party goes on until the last beer is finished, which usually doesn't take too long.[143]

GASLIGHT BREWERY AND
RESTAURANT | SOUTH ORANGE

More than sixteen years ago, the Soboti family decided that they all wanted to take on a project together. The matriarch, Cynthia, had long worked in the restaurant business and knew that she wanted to open up her own restaurant. Meanwhile, her husband, Dan Sr., and son, Dan Jr., were avid homebrewers who owned a homebrew supply store, and Dan Jr. had dreams of running their own brewery. There was one obvious answer: open a brewpub. And that's exactly what they did when they opened Gaslight Brewery and Restaurant in South Orange in June 1998. Even today, Gaslight is the only brewpub in Essex County.

Since opening, Dan has brewed his way into the hearts of local residents with beers like his 1920s Lager and Perfect Stout, brewed on the eight-barrel Pub Brewing system. Like many other brewpubs, Gaslight showcases its beautiful copper tanks right in front of the large window that faces out onto South Orange Avenue. Unlike most brewpubs, though, the pub also offers beers that aren't its own, both on tap and in bottles. Recent guest breweries have been Stone (California), Terrapin (Georgia), Tröegs (Pennsylvania) and Kane (New Jersey).[144]

Beer runs deep in the Soboti family. The father-and-son team had more than a decade of homebrew experience before opening up Gaslight, including owning a homebrew store—homebrewers can still buy all the equipment and ingredients they need at U-Brew Homebrewing, the supply store run by Dan Jr. that sits on top of Gaslight. Dan Jr. even completed the Masters Brewing Program at the University of California–Davis.

Gaslight is also host to the Draught Board 15, a local beer and homebrew club that allows members to trade recipes, sample one another's homebrews, share interesting beer they've found and, of course, sample the recent Gaslight offerings. It also teams up to send a bus to TAP New York Craft Beer and Food Festival every year.

Gaslight has also developed quite the reputation for catering to the special events and seasons. The brewpub brews a popular Oktoberfest every year that gets paired alongside an entire Oktoberfest food menu. It also hosts a large Victorian Christmas Dinner at which beers (especially Dan's Satan Claws Barleywine), wines and even meads are paired alongside a list of specially prepared food.[145]

KROGH'S RESTAURANT & BREW PUB | SPARTA

Many of New Jersey's beer producers have long, rich histories, but perhaps none more so than Krogh's Restaurant & Brew Pub in Sparta. The town was established in 1845 to serve as a central community among the farms and mountains that made up northwest New Jersey. In the early 1920s, the Arthur D. Crane Company bought thousands of acres of land known as Brogden Meadow, dammed the valley and flooded it, creating Lake Mohawk. Over time—thanks in large part to the lake—the area grew, and so did its population.

In 1927, a small teashop was opened, and following the end of Prohibition, it was converted into a tavern called the Carl Malmquist Restaurant. In 1937, a Dutch man named Frede Krogh bought the building and turned it into Krogh's Restaurant and Tap Room, which his family owned until 1973.[146]

In 1981, Robert "Bob" Fuchs, a Kearny-born entrepreneur, bought the building. When New Jersey laws changed to allow brewpubs, Bob quickly went to work. After three years of planning, financing and building, the brewery began operating in 1999, and the restaurant's name was changed yet again to Krogh's Restaurant & Brew Pub. It was the first brewpub in Sussex County.

Krogh's Brogden Meadow Pale Ale. *Julian Huarte.*

Over the years, Krogh's has developed a well-earned reputation for making some of the best food and beer in the state. Brewing on a five-barrel system with five fermenters and seven serving tanks, head brewer Dave Cooper has been brewing at capacity (five hundred barrels) for some time. Beers like Krogh's Gold, Alpine Red Glow Ale, Old Krogh Oatmeal Stout and Three Sisters Golden Wheat get their names "with local flavor," as Bob Fuchs said. Alpine Red Glow, for example, is named after the alpine section of Lake Mohawk, while Three Sisters Golden Wheat is named in honor of Dave's three sisters. Seasonal beers like the popular Oktoberfest are also added to the mix, and many get their names from loyal customers and employees. Blum Belgian Wit, for example, is named after Charles and Charlie Blum, who have been going to Krogh's for decades. Augie's Alt is named after the restaurant's manager, Karen (whose nickname is Augie), who has worked at Krogh's for more than thirty years.

The beer names pay homage to the close-knit community that is Krogh's; according to Fuchs, Krogh's employs more than fifty people, three-fourths of whom have been with him for more than twenty years. Longevity is important to Fuchs. Although the building that houses Krogh's is more than ninety years old, it has only had only four owners. "My goal is to hold the longest chapter and I am approaching that goal. The Kroghs owned Krogh's for thirty-seven years, and it is my goal to break that record...I am now entering my thirty-fourth year. Krogh's is a very, very special place."[147]

Indeed it is, and you can tell as soon as you approach the building. High-peaking roofs punctuate stone walls amid a mountainous backdrop. As Bryson and Haynie wrote, the building "conjure[s] up visions of fairy tales and storybook characters."[148]

Krogh's has done a handful of special beers to commemorate special events in the brewery's history. Most recently, to celebrate the brewery's fifteenth anniversary, a Russian imperial stout was conditioned in the Sterling Mines of Ogdensburg, New Jersey, for a year; this replicated the traditional methods of conditioning beer in the times before refrigeration. The brewery had to receive special permission from the state's Alcohol Beverage Control to transport and condition beer outside the brewpubs walls.

While the tradition of Krogh's will carry on for many years to come, Fuchs and his staff are beginning to prepare for the next chapter in the Krogh's story: a production brewery. According to Fuchs, the five-hundred-barrel capacity comes up several hundred barrels short of the demand for Krogh's beers. With a production brewery, Fuch's hopes to be producing ten thousand barrels within the next five years.[149]

CRICKET HILL BREWERY | FAIRFIELD

Rick Reed, founder of Cricket Hill Brewery in Fairfield, is one of the more passionate brewers not just in New Jersey but in the whole country. In 2008, he made noise throughout the craft beer world when a YouTube user posted a video he took during a tour of Cricket Hill. In the video, Reed stands up in front of a crowd of a few dozen tour-goers and delivers a passionate speech on craft beer and his brewery's place in the industry.

Speaking about the macrobrewers (specifically Budweiser, Coors and Miller), Reed said things like, "You have been brainwashed since you were children…to convince you that there is no reason to drink anything but those three beers, and to this day, we still haven't grasped the fact that they are lying to us, that they think we are stupid. The number-one selling water in New Jersey is Coors Light. How stupid do they think we are?" He also said, "Coors Light is the coldest-tasting beer…I'll tell you what it means: it means nothing…They tell us that Coors Light is frost brewed. That means nothing…It's brewed at 212 degrees just like every other beer."[150]

His passion against macro-produced beer is exceeded only by his passion for his own. "We have to taste what we drink, and that's why we're here,

Cricket Hill Hopnotic IPA. *Chris Morris.*

drinking the best beer ever made on the planet Earth: Cricket Hill...If you enjoy Cricket Hill, you drink Cricket Hill...Cricket Hill is the finest beer ever made on the planet Earth."[151]

That's a tall order to fill, but it's one Reed and his team have been working hard to fulfill since Cricket Hill formed in 2000 (although it didn't start brewing until 2002). The mission of the brewery has always been simple enough: to brew full-bodied but sessionable beers. His inspiration to start a brewery began as early as 1972, when he moved to New Jersey and had his first taste of Ballantine Ale. As Reed said, he wants Cricket Hill to serve as transitional beers so that drinkers of beers like Budweiser and Coors Light can be introduced to—and still enjoy—beers with more flavor.

With brewers Mark Tilley and Vinny Tamburri on a fifteen-barrel system alongside Cellerman Mark Ryan, Ed Gangi and, of course, Rick Reed, that's exactly what Cricket Hill is doing. The first Cricket Hill beers were East Coast Lager and American Ale, first brewed in 2002. Both serve as excellent examples of sessionable yet flavorful beers. Hopnotic IPA and Colonel Blides ESB joined the lineup over the next few years, and in 2009, the brewery began to brew several seasonals—including the unique Jersey Breakfast Ale, Nocturne Chocolate Ale, Fall Festivus and Pumpkin Ale. Those were all followed with the brewery's Reserve and Small Batch series—with beers like the Abbey Cherry Tripel, Jersey Devil Imperial Red and Bourbon Barrel Aged Imperial Porter—which offer a bit more flavor than the sessionable year-rounds and seasonals.[152]

CHAPTER 9
THE CRAFT BEER BOOM

2005–2014

There's no agreed-upon definition of exactly when the "craft beer boom" began. Really, craft beer came onto the scene as early as the mid-1960s, when Fred Huh recommended that Fritz Maytag take a trip to the Anchor Steam brewery, where his favorite beer, Anchor Steam, was brewed. As we know, Fritz ended up purchasing Anchor, and the craft beer revolution began. But just as one stock doesn't constitute a stock market "bubble," one brewery can't count as a craft beer "boom," can it? Perhaps, then, we need to fast-forward a few years to 1978, when New Albion Brewing became the first microbrewery of the modern era. Still, though, two breweries is hardly a revolution. Within five years, though, craft beer saw the likes of Sierra Nevada, Redhook and Bell's. The boom was becoming real.

Throughout the 1980s and early 1990s, the boom indeed became real. Famed breweries like Lagunitas, Firestone Walker, Boston Beer (Sam Adams), Brooklyn Brewery, Deschuetes, New Belgium and Dogfish Head opened around the country. But the craft beer revolution that was beginning to take hold in states like California, Massachusetts and even New York wasn't taking hold in New Jersey. Until 1994, the Garden State boasted just one brewery: Anheuser-Busch. Our revolution didn't begin until 1994, when Dave Hoffman started Climax.[153]

Throughout the 1990s, the Jersey beer scene began to take shape. In addition to Climax, there were production breweries like High Point and River Horse in the northern half of New Jersey, as well as Flying Fish in the southern half. Brewpubs like Long Valley, Harvest Moon, J.J. Bitting, Trap

Left: Lacing leaves a mark after every sip of Cricket Hill's Hopnotic IPA. *Chris Morris*.

Below: Jeremy "Flounder" Lees of Flounder Brewing gives a tour during Flounder's November 2013 grand opening. *Flounder Brewing*.

Rock, Pizzeria Uno, Gaslight and Krogh's were joined by Basil T's (Red Bank and Toms River), Artisan's (Toms River) and the Tun Tavern (Atlantic City). Had Cricket Hill not formed in 2000, the Garden State would have gone more than a decade with no new production breweries.

In 2007, two years after Egan & Sons brewpub formed in Montclair, Brian Boak brewed his first commercial beer, Monster Mash, under the BOAKS Beer banner. In North Jersey, BOAKS was followed by New Jersey Beer Co. in 2010, Bolero Snort and Flounder in 2013 and Angry Erik and 902 Brewing in 2014. After a long period of stability in Jersey craft beer, things were finally starting to get shaken up. And this doesn't include all the breweries that were popping up down the shore and in South Jersey, like Carton, Kane, Cape May, Tuckahoe, Beach Haus and Rinn Duin. Those are for another book.

EGAN & SONS | MONTCLAIR

The craft beer enthusiast might have a problem with Egan & Sons appearing in this book, since it doesn't brew in the traditional sense. The brewpub uses a Breworx system, where already made wort is fermented and carbonated before being served. When I started homebrewing many years ago, I did what's called extract brewing, which is the way most homebrewers get started. Instead of taking milled grains and mashing them to create sugar, you use malt extract and add it to boiling water. Then add your hops, cool it, ferment it, carbonate it and drink it. Sure, it's a bit more involved than Egan & Sons, but not by much. At the very least, it creates and serves unique beers and deserve an honorable mention.

Egan & Sons is the creation of Chris Egan, a man with plenty of experience in craft beer. Hailing from Dublin, Egan grew up working in pubs throughout Europe, at one time owning four pubs in Dublin. He even worked at a brewery, Commonwealth Brewing in Boston, and this made him want to start his own. But the investment and space that Commonwealth demanded was a bit much, which is why Egan chose to use a Breworx system for his pub. Nonetheless, four house beers are always available on tap at Egan's, and that's certainly better than none.[154]

BOAKS BEER | POMPTON LAKES

Brian Boak has long been brewing some of the best beer in the state of New Jersey. He claims that his mother called him a beer snob in 1977, before being a beer snob was even a thing; he was drinking imported beer while everyone else drank cheap macro beer. He didn't begin brewing until 1999, when his kids bought him a homebrew kit. Within a few years, he was a homebrewer, winning gold at a beer festival in 2002 with his Monster Mash Imperial Stout. He hit the big time when that same beer won Best in Show at the New Jersey State Fair.

In December 2007, Monster Mash became Brian Boak's first commercial produced beer, brewed under the banner of BOAKS. A few months later, it was joined by Two Blind Monks, and both went on sale to the public in February 2008. Later that year, Abbey Brown Ale joined the lineup, followed by Double BW in 2009 and Wooden Beanie (with the Abbey Brown aged in Jack Daniels barrels and vanilla beans) in 2010. As of 2014, BOAKS beers are contract-brewed at High Point, and Greg Zaccardi's brewery is able to accommodate the increased output thanks to the fermentation tank purchased by BOAKS.

Boak in undoubtedly proud of his award-winning background; it is, after all, what put BOAKS on the map and convinced him to go pro. But

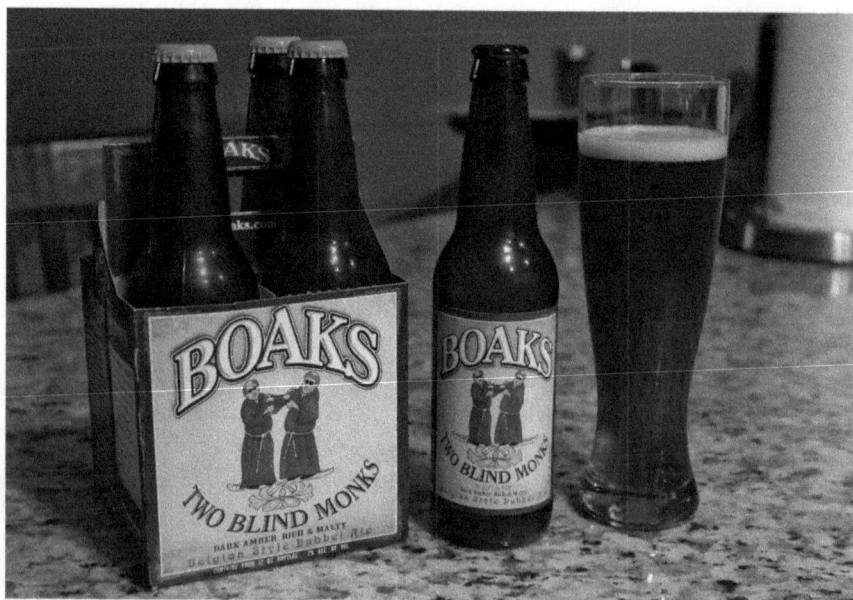

BOAKS' Two Blind Monks Belgian Dubbel Ale. *Chris Morris.*

he's quick to explain that his beers aren't necessarily the type that will go on to win dozens of awards: "BOAKS beers are brewed to taste, not strict specifications. Strict specifications win Gold Medals; brewing great tasting beers wins fans."[155]

NEW JERSEY BEER CO. | NORTH BERGEN

If you couldn't tell from the name, New Jersey Beer Co. is all about the Garden State. It's where the brewery's chairman, manager, brewers, staff and investors were all born and raised. They take a lot of pride in that, and they should.

New Jersey Beer Co. was founded in 2010 by Matt Steinberg—who lived in North Bergen at the time—after years of homebrewing. Along with brewer Brendan O'Neil and a small staff, New Jersey Beer Co. was launched, but not without its problems. The brewery's bottling line broke before filling its 700th case, posing an obvious problem for getting New Jersey Beer Co. beers in the homes of the state's residents. The brewery found itself in need of a financial rescue.[156]

Enter Paul Silverman, a real estate developer in neighboring Jersey City. Silverman had long wanted to open his own brewery—Jersey City Brewing—and around 2011, he approached Steinberg looking for advice on how to do it. Steinberg made him a different offer: join New Jersey Beer Co. instead.

It took a few weeks for Silverman to pull the trigger. Silverman is the owner of SILVERMAN, a real estate development company that he runs with his brother, Eric. For decades, they had taken older properties, gutted them and started from scratch. It was their way of making sure that things were done right, rather than rely on, say, the plumbing that had previously been done. Silverman wanted to take the same approach to his brewery: he wanted to build it himself the way he wanted it. After weeks of thinking over Steinberg's offer, something—maybe fate—happened: Silverman saw John Holl, a well-known beer writer, on *The Today Show*. Holl, it turns out, lived in Jersey City, the place Silverman had dedicated so much of his life to building. During the segment, Holl mentioned Garden State Stout, a beer brewed by New Jersey Beer Co. The next day, Silverman called Steinberg. He wanted to join New Jersey Beer Co.

Steinberg has since moved on to other projects, and Silverman has grown more involved, becoming chairman of the company. A lot of other things have changed since then, too.

The brewery, which started with four twenty-barrel fermenters, has doubled and currently brews about three thousand barrels per year, with room to grow. Its product line has grown in similar fashion. When the first New Jersey Beer Co. beer was tapped at the Iron Monkey in May 2010, the brewery offered three beers: Garden State Stout, Hudson Pale Ale and 1787 Abbey Single. Together, the lineup put an emphasis on the three main ingredients (not including water): malt, hops and yeast, respectively. They also paid tribute to the country and state in which they're brewed. Garden State Stout, of course, honors the state of New Jersey, while Hudson Pale Ale pays homage to Henry Hudson—an early explorer of the New York area—and 1787 Abbey Single honors the year New Jersey became the third state of the new union.

The brewery has since introduced LBIPA, a 7 percent ABV East Coast IPA inspired by the Jersey Shore (LBI is a reference to Long Beach Island, a popular destination on the Jersey shore), with a portion of all sales going to the Alliance for a Living Ocean (www.livingocean.org). LBIPA received praise from Charlie Papazian, president of the Brewers Association, and recently won a silver medal at the 2014 Atlantic City Beer and Music Festival, the biggest craft beer festival in New Jersey. New Jersey Beer Co.'s other medal-winner, Weehawken Wee Heavy, took home bronze in 2013.

The brewery distributes through Allied Beverage Group, and its beers are served throughout New Jersey. In September 2014, the brewery took another giant step: it was put on thirty taps at Newark's Prudential Center, the home of the New Jersey Devils and Seton Hall basketball team.

Silverman attributed the success of the brewery to a few factors. He's quick to point out that his brewers are special: Brendan O'Neil has taken part in every batch of beer the brewery opened, and Dave Manka joined the team after six years of brewing for the popular Cricket Hill. He also emphasized the way their beers are made. Just as he likes to make sure his developments are done right in every aspect, the brewery places the same emphasis on quality. The malts, for example, are all German—more expensive but high in quality. Finally, he noted that much of the success of New Jersey Beer Co. comes from the people for whom the beer is brewed. "We are New Jersey. Our people, our beers, our story. We care about the beer we brew, and we love that our fans love our beer."[157]

BOLERO SNORT BREWERY | RIDGEFIELD PARK

Bob Olson first brewed a partial mash Summer Ale in June 2009. A friend of his had taken on homebrewing, and Bob thought that it might be a fun hobby for him to do, too. In April 2010, he moved to all-grain brewing. By January 2013, he was a professional brewer, the founder of Bolero Snort.[158]

The name is intriguing, I know. Olson explained that as a homebrewer, he used to label each of his beers with a picture of himself. Everyone who tried his beer loved it, but they weren't all so fond of the label. Trying to keep his identity in the label, Olson entered his full name, Robert Olson, into an anagram generator. That's where Bolero Snort was born. He designed a logo with a bull on it and came up with the brewery's clever motto: "No BS, just ragin' good beer."

Along with co-founder and head brewer Andrew Maiorana, Olson and Bolero Snort have been gaining notoriety not only in North Jersey—where they distribute most of their beer (for now)—but also throughout the state and beyond. The list of award-winning beers is almost as long as the entire list of beers and includes their flagship beers like Ragin' Bull Longhop IPA and There's No Rye-ing in Basebull, seasonals such as Lucky Buck and Gingerbull Cookie and limited-release beers like Happy Buck'n Anniversary. The brewery even won People's Choice at the 2014 Atlantic City Beer and Music Festival—the biggest craft beer festival in New Jersey.

Bolero Snort is currently contract-brewed at High Point, which it's able to do thanks to the fermenter it had specially installed just for Bolero. It is currently planning a move to its own location.[159]

FLOUNDER BREWING COMPANY | HILLSBOROUGH

In 2005, Jeremy "Flounder" Lees turned his tiny Morristown kitchen into a small homebrewery, where he began crafting beers that he thought others would enjoy. After bringing his brothers, Dan and Mike, and cousin, Will, in on the brewing, the three of them upgraded to a SABCO brewing system, forcing them out of Jeremy's kitchen and into their grandmother's Lyndhurst garage. Their new home became a weekend destination, where friends and family came to barbecue, sit around a fire and, of course, enjoy the brothers' fine brews.

By 2007, they knew that they wanted to take the next step. Not wanting to quit their day jobs, they opted on a small nanobrewery, which they could

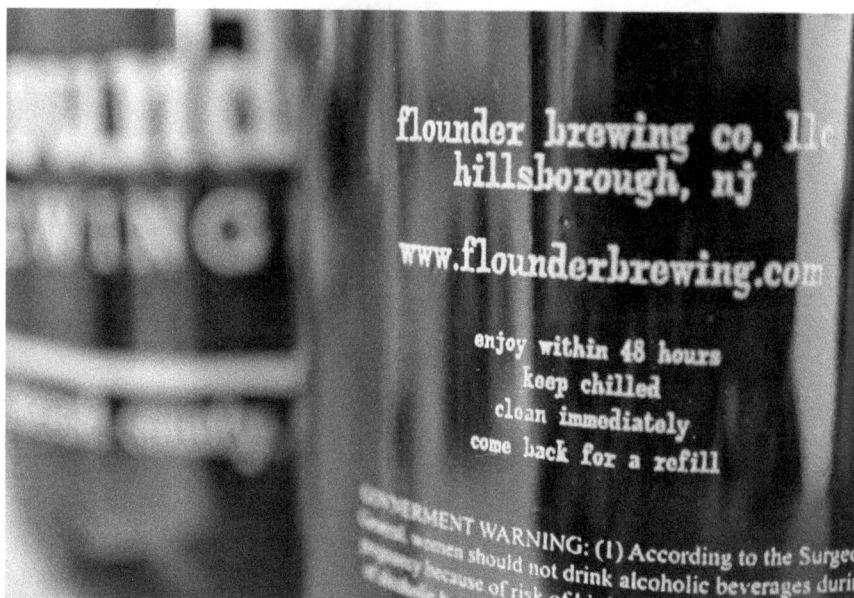

Growlers at Hillsborough's Flounder Brewing advise you to drink your beer quickly. *Flounder Brewing.*

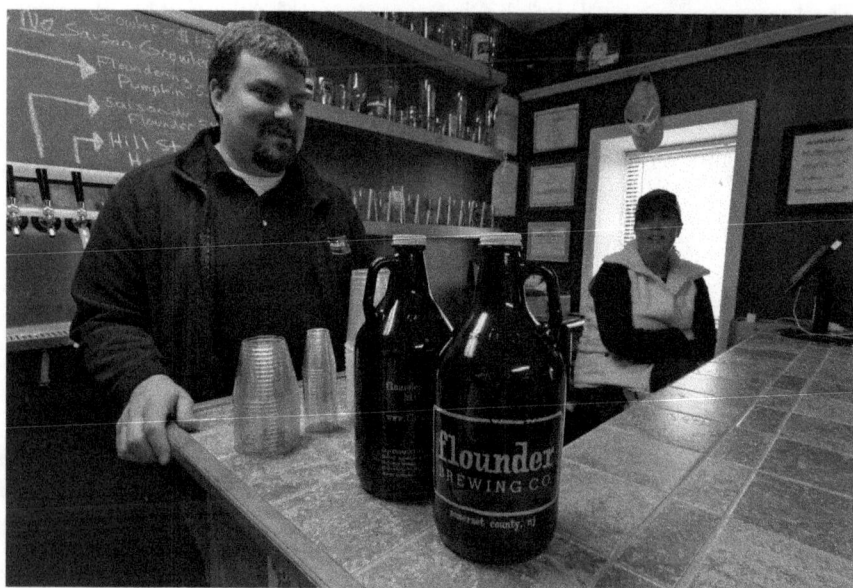

Jeremy "Flounder" Lees of Flounder Brewing stands with two growlers in the taproom. *Flounder Brewing.*

always expand. In 2010, they signed a lease for their building in Hillsborough. It took several years of maneuvering through New Jersey's complicated legal landscape, but in October 2013, the brewery finally produced its first batch of beer: Hill Street Honey Ale. They have since introduced two more flagships, Saison du Flounder and Murky Brown Ale, and currently produce about fifty barrels per year.[160]

ANGRY ERIK BREWING | LAFAYETTE

The year 2014 saw two breweries form in North Jersey. The first was Erik and Heide Hassing's Angry Erik Brewing. Based in Lafayette, Angry Erik currently consists of a ten-barrel brew house with two fermenters and a bright tank. Its first two beers were Ravol, an American amber ale, and 3 Ball Porter, brewed with cardamom and orange. Promising an "ever-changing variety," Angry Erik has followed those with beers such as the Dainty Viking, a blonde ale brewed with elderflower; Mary Shroom Winter Saison; Hop the Fence IPA; Hop-N-Awe Double IPA; and Kayewla Belgian Pale Ale.[161]

902 BREWING COMPANY | HOBOKEN

The story of beer in New Jersey started in 1641 with Aert Teunissen Van Patten in Hoboken. It's only fitting that for now (emphasis on the "for now"), the story ends back in the Mile Square City. 902 Brewing started at 902 Washington Street in Hoboken a decade ago, when roommates Tucker Littleton and Colby Janisch began homebrewing together. The two, along with friend Andrew Brown, took it to the next level when they began planning 902 Brewing Company in 2012. In November 2014, 902 became a legally licensed brewery. The 902 trio is currently producing small-sized batches of Pale Ale True Hoboken (PATH), Black Dynamite Black IPA and Inconsiderate Dawwwg Stout as they expand the brewery.[162]

BEER OUTSIDE OF BREWERIES

While New Jersey was once a brewing mecca, the last decades have seen the Garden State fall far behind in terms of craft breweries. But that's not to say that New Jerseyans haven't been enjoying their craft beer at all. Over the past two decades, craft beer bars have offered New Jersey's biggest beer fans some of the finest beers in the world. Bars like Barcade (Jersey City), World of Beer (New Brunswick), the Copper Mine Pub (North Arlington), Morris Tap & Grill (Randolph), the Iron Monkey (Jersey City), Pilsener Haus Biergarten (Hoboken), Zeppelin Hall Biergarten (Jersey City), Stuff Yer Face (New Brunswick) and the Tap Room offer dozens of craft selections on tap. The Cloverleaf Tavern in Caldwell has twice been named the best craft beer bar in the northeast United States in Craftbeer.com's Great American Beer Bars competition.

These bars and others offer more than just dozens of taps or hundreds of bottles, though. Many have earned reputations for things like beer dinners—pairing a specially crafted food menu with carefully chosen beers—tap takeovers or special brewery nights. Hailey's Harp & Pub in Metuchen even has what it calls a Beer Senate. Once a month, beer guru Moshe Atzbi holds what is essentially a beer dinner with a twist.

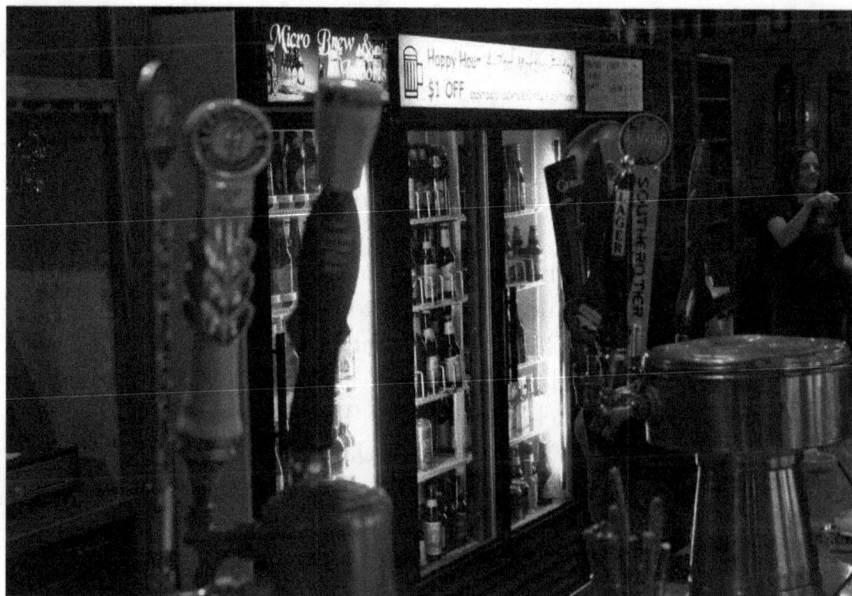

New Brunswick's Stuff Yer Face offers an impressive tap and bottle offering. *Chris Morris.*

Cloverleaf Tavern in Caldwell has been voted best bar in the Northeast multiple times. *Cloverleaf Tavern.*

Rather than simply enjoy the specially cooked food and hand-selected craft beers, Beer Senators, as they're known, rate the beers and vote on their favorites. The winner is put on tap the next month as part of the rotating tap selection—it's beerocracy at its finest. Following the dinner, spirited discussion ensues, such as debates on bottle versus can versus draft versus cask.

One final noteworthy non-brewery is New Jersey Craft Beer, an organization aimed at promoting the craft beer scene in New Jersey. From news to events to its membership club—where members save money when they purchase craft beer from participating retailers and restaurants—New Jersey Craft Beer and President Mike Kivowitz are dedicated to promoting anything and everything craft beer in the state of New Jersey. I imagine they'll be pretty busy for some time to come.

CHAPTER 10

THE CURRENT STATE OF BEER IN NORTHERN JERSEY

Without a doubt, the craft beer industry and culture is thriving in New Jersey, as Jerseyans are putting down cans of Budweiser and Coors Light in favor of real, flavorful and fresh beer, much of it made right in their own neighborhoods. The Garden State was once a mecca for brewing, but Prohibition put an end to that era. Over the next thirteen years, most Jersey breweries shut down, while some produced near beer and others produced illegal beer under the operations of some of the most famous gangsters of all time like Abner "Longy" Zwillman, Max Hassel, Irving "Waxey Gordon" Wexler and Max "Big Maxie" Greenberg.

With the end of Prohibition came a glimmer of hope. Breweries like Krueger and Ballantine were once again nationally recognized names, and they played a part in developing some of the innovations that are now commonplace today, like canned beer. But this new life was short-lived. Following World War II, a decades-long period saw the large breweries consolidate, and even the biggest of Jersey breweries could no longer compete. By 1984, there was only one brewery remaining: Anheuser-Busch. For decades, not a single craft beer was brewed in New Jersey.

That all began to change in June 1994, when Dave Hoffman opened up Climax in the back of his father's store. He was followed by the likes of High Point, River Horse, Cricket Hill, BOAKS, New Jersey Beer Company, Bolero Snort, Flounder, Angry Erik and 902 Brewing, while brewpubs like the Ship Inn, Triumph, Long Valley, Harvest Moon, J.J. Bitting, Trap Rock, Pizzeria Uno, Krogh's and Egan & Sons provided their beer to in-house customers.

In May 1994, there wasn't a single craft brewer in the state of New Jersey. Today, there are forty, with half of them in the north.

And it doesn't appear that things will slow down anytime soon. There are currently more than twenty-five breweries in the planning stages, with names like Black Dog Brewing (Metuchen), Demented Brewing (Middlesex), Departed Soles Brewing (Jersey City), Conclave Brewing (Raritan), Cypress Brewing (Edison), Hoboken Brewing (Hoboken), Miscin Brewing (Lafayette), Stibinger Brother Beer (Sparta) and Wet Ticket Brewing (Rahway), all in the north.

Notes

Chapter 1

1. Paul S. Boyer, *The Oxford Companion to United States History* (New York: Oxford University Press, 2001), 87.
2. Charlie Papazian, "The Revival of American Beer," Craftbeer.com, accessed October 10, 2014.
3. Tom Acitelli, *The Audacity of Hops: The History of America's Craft Beer Revolution* (Chicago: Chicago Review Press, 2013), 4.
4. Jack Lefler, "Steam Beer Is Hot Item Among Young in Frisco," *Milwaukee Journal*, June 9, 1975, 23, accessed October 10, 2014.
5. Acitelli, *Audacity of Hops*.
6. Brewers Association, "U.S. Brewery Count Tops 3,000," July 9, 2014, http://www.brewersassociation.org/insights/us-brewery-count-tops-3000, accessed October 11, 2014.
7. Wynkoop Brewing Company, "Rocky Mountain Oyster Stout," July 2, 2013, http://www.wynkoop.com/blog/wynkoop-releases-first-cans-of-rocky-mountain-oyster-stout, accessed October 11, 2014.
8. Carton Brewing Company, http://cartonbrewing.com/#!/panzanella, accessed October 11, 2014.

Chapter 2

9. Beer Institute, "Beer History," accessed October 11, 2014; Martin Isler, *Sticks, Stones, and Shadows: Building the Egyptian Pyramids* (Norman: University of Oklahoma Press, 2001), 265.

10. John J. Palmer, *How to Brew: Everything You Need to Know to Brew Beer Right the First Time*, 3rd ed. (Boulder, CO: Brewers Publications, 2006).

11. David Martorana, "The Short and Bitter History of Hops," Beer Scene, May 1, 2010, 2014, http://www.beerscenemag.com/2010/04/the-short-and-bitter-history-of-hops, accessed October 11.

12. Palmer, *How to Brew*.

13. Abigail Tucker, "The Beer Archaeologist," *Smithsonian Magazine* (August 1, 2011), accessed October 13, 2014

14. Palmer, *How to Brew*.

Chapter 3

15. Michael Pellegrino, *Jersey Brew: The Story of Beer in New Jersey* (New Jersey: Lake Neepaulin Publishing, 2009), 4.

16. *One Hundred Years of Brewing* (New York: H.S. Rich & Co., 1903), 140; James Deetz and Patricia Scott Deetz, *The Times of Their Lives: Life, Love, and Death in Plymouth Colony* (New York: Random House, 2001), 8; Nathaniel Philbrick, *Mayflower: A Story of Courage, Community, and War* (New York: Penguin Group, 2006), 79.

17. *One Hundred Years of Brewing*, 140; William Nelson, *Documents Relating to the Colonial History of the State of New Jersey*, vol. 19 (Paterson, NJ: Press Printing and Publishing, 1897), 208, 408.

18. Beer Institute, "Beer History," accessed October 11, 2014.

19. Pellegrino, *Jersey Brew*, 6–7.

20. Amy Mittelman, *Brewing Battles: A History of American Beer* (New York: Algora Publishing, 2008), 88–98.

21. Ibid., 13–14.

Chapter 4

22. Jim Hans, *100 Hoboken Firsts* (Hoboken, NJ: Hoboken Historical Museum, 2005); Sarah E. Van Patten Ellsworth and Elizabeth Young, *Genealogical Record of the Van Putte, Petten, Patten Family* (N.p., 1915).

23. Lew Bryson and Mark Haynie, *New Jersey Breweries* (Mechanicsburg, PA: Stackpole Books, 2008), xv–xvii; Brewers Association, "Craft Beer Sales by State," accessed October 16, 2014.

24. *Albany Evening Journal*, "Valuable Brewery at Auction," January 1, 1831, accessed October 21, 2014.

25. Jean-Rae Turner and Richard T. Koles, *Newark, New Jersey* (Charleston, SC: Arcadia Publishing, 2001), 45; Bryson and Haynie, *New Jersey Breweries*, xvi.

26. Bryson and Haynie, *New Jersey Breweries*; *New York Times*, "Other Deaths: Peter Ballantine," January 24, 1883, accessed October 21, 2014.

27. *New York Times*, "Other Deaths: Peter Ballantine."

28. Francis Bazley Lee, ed., *Genealogical and Memorial History of the State of New Jersey* (New York: Lewis Historical Publishing Company, 1910), 1,065; Bryson and Haynie, *New Jersey Breweries*, xvi.

29. Frederick William Salem, *Beer, Its History and Its Economic Value as a National Beverage* (Hartford, CT: F.W. Salem & Company, 1880), 183–84; Frank J. Urquhart, *A History of the City of Newark, New Jersey*, vol. 2 (New York: Lewis Historical Publishing, 1913), 1091.

30. William S. Myers, *Prominent Families of New Jersey* (Baltimore, MD: Genealogical Publishing, 2000), 361.

31. Pellegrino, *Jersey Brew*, 20.

32. *Newark, the City of Industry: Facts and Figures Concerning the Metropolis of New Jersey* (Newark, NJ: Newark Board of Trade, 1912), 169; *Western Brewer and Journal of the Barley Malt and Hop Trades*, vol. 33 (New York: H.S. Rich & Co., 1908), 592; vol. 52, 199.

33. Old Newark, "Old Newark Business & Industry: Bottlers, Brewers & Wineries," 615, http://oldnewark.com/mainindex.php, accessed October 30, 2014; Myers, *Prominent Families of New Jersey*, 361; *New York Times*, "C.W. Feigenspan, Newark Brewer; President of Company Dies at His Rumson Home," February 7, 1939, accessed November 4, 2014.

34. *New York Times*, "C.W. Feigenspan, Newark Brewer."

35. Carmela Karnoutsos, "Lembeck & Betz Eagle Brewing Company," Jersey City Past and Present, New Jersey City University, https://www.njcu.edu/programs/jchistory/Pages/L_Pages/Lembeck.html, accessed November 2, 2014.

36. Mark A. Noon, *Yuengling: A History of America's Oldest Brewery* (Jefferson, NC: McFarland & Company, 2005), 73–75.

37. Karnoutsos, "Lembeck & Betz Eagle Brewing Company"; Pellegrino, *Jersey Brew*, 30.

38. Pellegrino, *Jersey Brew*, 50; William E. Sackett, ed., *Scannell's New Jersey's First Citizens and State Guide* (Paterson, NJ: J.J. Scannell, 1918), 237–38.

39. Gregg Smith, "The Doelger Breweries," Real Beer, http://www.realbeer.com/library/authors/smith-g/doelger_ny.php, accessed November 3, 2014; Old Breweries, "Peter Hauck Brewery," www.oldbreweries.com, accessed November 2, 2014.

40. T.F. Fitzgerald, *Manual of the Legislature of New Jersey*, vol. 114 (Trenton, NJ: self-published, 1890), 222; *New York Times*, "Rivers of Jersey Beer," December 3, 1889, accessed October 27, 2014.

41. *New York Times*, "Rivers of Jersey Beer."

42. Jean-Rae Turner, Richard T. Koles and Charles F. Cummings, *Newark: The Golden Age* (Columbia, SC: Arcadia Publishing, 2003), 29.

43. Pellegrino, *Jersey Brew*, 23–53; *New York Times*, "C.W. Feigenspan, Newark Brewer."

44. Wolfgang Saxon, "G.E. Wiedenmayer, Company Chairman and Ex-Banker, 85," *New York Times*, October 21, 1993, accessed November 2, 2014; *Western Brewer and Journal of the Barley Malt and Hop Trades*, vol. 34, 473–74.

45. *Western Brewer and Journal of the Barley Malt and Hop Trades*, vol. 34, 473–74.

46. Saxon, "G.E. Wiedenmayer, Company Chairman."

47. Jean-Rae Turner and Richard T. Koles, *Elizabeth: First Capital of New Jersey* (Charleston, SC: Arcadia Publishing, 2003), 92.

48. Ibid.; Carl W. Schlegel, *Schlegel's American Families of German Ancestry* (New York: Genealogical Publishing, 2003), 138–40.

49. Charles A. Shriner, *Paterson, New Jersey: Its Advantages for Manufacturing and Residence* (Paterson, NJ: Press Printing and Publishing Company, 1890), 212–14.

50. Ibid., 211–12; Old Breweries, "Hinchliffe's Brewery," accessed November 2, 2014.

51. Pellegrino, *Jersey Brew*, 11.

52. Ibid., 29; New Jersey State Library, "Corporations of New Jersey," 143, accessed November 8, 2014.

53. Turner and Koles, *Elizabeth*, 92; Pellegrino, *Jersey Brew*, 77.

54. Pellegrino, *Jersey Brew*, 34–36.

55. Old Breweries, accessed November 15, 2014.

Chapter 5

56. Noon, *Yuengling*, 103.

57. *Truth*, "Saloons, Sunday Laws and License Fees," August 1, 1903, 7.

58. Pellegrino, *Jersey Brew*, 69–71.

59. Rich Wagner, *Philadelphia Beer: A Heady History of Brewing in the Cradle of Liberty* (Charleston, SC: The History Press, 2012), 89.

60. Pellegrino, *Jersey Brew*, 74–86.

61. Falstaff Brewing, "Ballantine XXX Ale," February 20, 2012,, http://www.realbeer. com/library/authors/smith-g/doelger_ny.php, accessed October 26, 2014.

62. Pellegrino, *Jersey Brew*, 19–50.

63. Ibid., 76.

64. Ibid., 82; David E. Kyvig, *Repealing National Prohibition* (Chicago: University of Chicago Press, 1979), 30.

65. Marc Mappen, *Jerseyana: The Underside of New Jersey History* (New Brunswick, NJ: Rutgers University Press, 1992), 176; Pellegrino, *Jersey Brew*, 77.

66. Pellegrino, *Jersey Brew*, 78; Mappen, *Jerseyana*, 176–77; Colonel Ira L. Reeves, *Ol' Rum River* (Chicago: Thomas S. Rockwell Company, 1931), xviii.

67. Reeves, *Ol' Rum River*, viii–ix.

68. Kevin Gibson, *Louisville Beer: Derby City History on Draft* (Charleston, SC: The History Press, 2014), 79; Pellegrino, *Jersey Brew*, 82.

69. Pellegrino, *Jersey Brew*, 83; David G. Moyer, *American Breweries of the Past* (Bloomington, IN: AuthorHouse, 2009), 7.

Chapter 6

70. Pellegrino, *Jersey Brew*, 85–87.

71. Mike Dash, *The First Family: Terror, Extortion, Revenge, Murder, and the Birth of the American Mafia* (New York: Random House Publishing, 2009), 268.

72. Clint Willis, ed., *Wise Guys: Stories of Mobsters from Jersey to Vegas* (New York: Thunder's Mouth Press, 2003), 143; William Poundstone, *Fortune's Formula: The Untold Story of the Scientific Betting System that Beat the Casinos and Wall Street* (New York: Hill and Wang, 2005), 31.

73. Poundstone, *Fortune's Formula*.

74. Pellegrino, *Jersey Brew*, 90.

75. Willis, *Wise Guys*, 144.

76. Pellegrino, *Jersey Brew*, 91–92.

77. Marc Mappen, *Prohibition Gangsters: The Rise and Fall of a Bad Generation* (New Brunswick, NJ: Rutgers University Press, 2013), 41–43, 174.

78. Ibid., 172; Pellegrino, *Jersey Brew*, 91–92.

79. Brad R. Tuttle, *How Newark Became Newark: The Rise, Fall, and Rebirth of an American City* (New Brunswick, NJ: Rutgers University Press, 2009), 100–101.

80. Pellegrino, *Jersey Brew*, 101–2.

81. Mappen, *Prohibition Gangsters*, 149; Poundstone, *Fortune's Formula*, 32.

82. Poundstone, *Fortune's Formula*, 35.

83. Ibid., 36; Mappen, *Prohibition Gangsters*, 150.

84. Mappen, *Prohibition Gangsters*, 149–51.

85. Edward A. Taggert, *Bootlegger: Max Hassel, the Millionaire Newsboy* (New York: Writer's Showcase, 2003), 11–12.

86. See Taggert, *Bootlegger*, 1–155.

87. Mappen, *Prohibition Gangsters*, 141–42.

88. Ibid., 32–34.

89. Walter Ehrlich, *Zion in the Valley: The Jewish Community of St. Louis* (Columbia: University of Missouri Press, 2002), 52.

90. Mappen, *Prohibition Gangsters*, 35.

91. Taggert, *Bootlegger*, 130–59.

92. Ibid., 156–60.

93. Ibid., 156–221.

94. Ibid., 223–26; Mappen, *Prohibition Gangsters*, 149.

95. Taggert, *Bootlegger*, 228–36.

CHAPTER 7

96. Bryson and Haynie, *New Jersey Breweries*, xvii.

97. Moyer, *American Breweries of the Past*, 7; Pellegrino, *Jersey Brew*, 82.

98. Pellegrino, *Jersey Brew*, 105–8.

99. Bryson and Haynie, *New Jersey Breweries*, xvii.

100. Ibid., xviii–xix; Pellegrino, *Jersey Brew*, 20–22.

101. Pellegrino, *Jersey Brew*, 15; Falstaff Brewing, "Ballantine XXX Ale," February 20, 2012, accessed October 26, 2014.

102. Pellegrino, *Jersey Brew*, 16–17.

103. Falstaff Brewing, "Ballantine XXX Ale"; Jeff Alworth "Rebuilding a Legend: Ballantine IPA," All About Beer, October 3, 2014, http://allaboutbeer.com/ballantine-ipa, accessed November 22, 2014.

104. Alworth "Rebuilding a Legend: Ballantine IPA."

105. Pellegrino, *Jersey Brew*, 50–51.

106. Bryson and Haynie, *New Jersey Breweries*, xix.

107. Jonathan Miller, "For 75 Years, It Was a Sight to Steer By in Newark," *New York Times*, June 27, 2006, accessed November 23, 2014.

108. Bryson and Haynie, *New Jersey Breweries*, xviii, 1–6.

109. Anheuser-Busch, "History," http://anheuser-busch.com/index.php/our-heritage/history, accessed November 22, 2014.

CHAPTER 8

110. Bryson and Haynie, *New Jersey Breweries*, xix–xxi.

111. Ibid., 14–15.

112. *The Brewery Show*, episode 008, "Climax Brewing," Blip Networks Inc., 2011.

113. Climax Brewing Company, www.climaxbrewing.com, accessed December 4, 2014; Bryson and Haynie, *New Jersey Breweries*, 15.

114. Bryson and Haynie, *New Jersey Breweries*,

115. *Brewery Show*, episode 008, "Climax Brewing"; Pellegrino, *Jersey Brew*, 124.

116. Bryson and Haynie, *New Jersey Breweries*, 113.

117. Ibid., 114.

118. Pellegrino, *Jersey Brew*, 131.

119. Bryson and Haynie, *New Jersey Breweries*, 113.

120. Ramstein Beer, "Ramstein: Traditional Lagers and Weizen Beers," http://www.ramsteinbeer.com, accessed December 4, 2014.

121. Bryson and Haynie, *New Jersey Breweries*, 114.

122. Ramstein Beer "Ramstein: Traditional Lagers and Weizen Beers," http://www.ramsteinbeer.com, accessed December 4, 2014; Bryson and Haynie, *New Jersey Breweries*, 93.

123. Bryson and Haynie, *New Jersey Breweries*.

124. Ibid., 96–97.

125. Ibid.; Ship Inn, http://www.britishbrewpub.com, accessed December 5, 2014.

126. Bryson and Haynie, *New Jersey Breweries*, 98–100.

127. Ibid., 124.

128. Great American Beer Festival, "GABF Winners," http://www.greatamericanbeerfestival.com/the-competition/winners, accessed December 8, 2014.

129. Bryson and Haynie, *New Jersey Breweries*, 125.

130. Ibid., 89.

131. Ibid.; Chris Ferullo, "River Horse Brewing Owner Talks Acquisition, Constraints and Craft Segment," Beer Pulse, June 19, 2012, http://beerpulse.com/2012/06/river-horse-brewing-owner-talks-acquisition-constraints-and-craft-segment, accessed December 8, 2014.

132. Bryson and Haynie, *New Jersey Breweries*, 90–91.

133. Interview with Chris Walsh and Dylan Bamrick, November 8, 2014.

134. NJ.com, "Glimpse of History: New Brunswick Brewing Company Hops in Late 1930s," April 10, 2011, accessed December 8, 2014.

135. Bryson and Haynie, *New Jersey Breweries*, 32.

136. Ben Bakelaar, "Interview with Harvest Moon Brewery," Love2Brew, http://www.love2brew.com/Articles.asp?ID=483, accessed December 9, 2014.

137. Harvest Moon Brewery & Cafe, http://www.harvestmoonbrewery.com, accessed December 9, 2014.

138. Bryson and Haynie, *New Jersey Breweries*, 37–39.

139. J.J. Bitting Brewing Co, http://www.njbrewpubs.com, accessed December 9, 2014.

140. Lloyd Nelson, "Bottoms Up! Cheer and Beer at Woodbridge's 7th Annual Central Jersey Beerfest," NJ.com, accessed December 9, 2014.

141. Bryson and Haynie, *New Jersey Breweries*, 45; Trap Rock Restaurant & Brewery, http://traprockrestaurant.net, accessed December 9, 2014.

142. Trap Rock Restaurant & Brewery website.

143. Bryson and Haynie, *New Jersey Breweries*, 41–43.

144. Ibid., 28; Gaslight Brewery & Restaurant, http://www.gaslightbrewery.net, accessed December 10, 2014.

145. Bryson and Haynie, *New Jersey Breweries*, 28–29.

146. Ibid., 118–19.

147. Interview with Robert Fuchs, November 30, 2014.

148. Bryson and Haynie, *New Jersey Breweries*, 118.

149. Interview with Robert Fuchs, November 30, 2014.

150. YouTube, "Cricket Hill Micro Brewery Plant Tour Speech," https://www.youtube.com/watch?v=91DH4lNpniE, accessed December 10, 2014.

151. Ibid.

152. Interview with Rick Reed, December 22, 2014.

CHAPTER 9

153. Acitelli, *Audacity of Hops*.

154. Bryson and Haynie, *New Jersey Breweries*, 22–25.

155. BOAKS Beer, http://boaks.com, accessed December 10, 2014.

156. Maggie Hoffman, "A Pint With: Matt Steinberg, New Jersey Beer Company," Serious Eats, August 12, 2010, http://drinks.seriouseats.com/2010/08/a-pint-with-matt-steinberg-new-jersey-beer-company.html, accessed December 5, 2014.

157. Interview with Paul Silverman, December 8, 2014.

158. From My Mellin, "Q&A with Homebrew Robert Olson of Bolero Snort," May 1, 2011, http://drinks.seriouseats.com/2010/08/a-pint-with-matt-steinberg-new-jersey-beer-company.html, accessed December 10, 2014.

159. Bolero Snort Brewery, www.bolerosnort.com, accessed December 11, 2014.

160. Flounder Brewing Company, http://www.flounderbrewing.com/age.php, accessed December 11, 2014.

161. Brian Casse, "Angry Erik Brewing Company," I Drink Good Beer, March 3, 2014, http://www.idrinkgoodbeer.com/craft-beer-blog/brewery-angry-eriks-brewing-lafeyette-jersey, accessed December 11, 2014; Angry Erik Brewing, http://www.angryerik.com, accessed December 11, 2014.
162. Steven Rodas, "Hoboken Microbrewery Gaining Momentum, Opens Storage Facility in Union City," NJ.com, accessed December 11, 2014.

INDEX

INDEX

About the Author

C hris Morris was born and raised in Metuchen, New Jersey, and has lived his entire life in the Garden State. He holds a BA from The College of New Jersey and an MA from Rutgers University and works full time on Wall Street. He brewed his first beer when he was twenty years old and was immediately hooked. He made a website, started a blog and even registered his brewery, Black Dog Brewing Company. He took his blogging to the next level when he started writing for the *Star-Ledger*, New Jersey's largest newspaper. He hopes to start his own brewery one day, but until then, he's enjoying exploring the craft beer world, as well as writing about it as he goes.

www.ingramcontent.com/pod-product-compliance
Lightning Source LLC
Chambersburg PA
CBHW070343100426
42812CB00005B/1412